Crawfish Farming for Beginners
Step-by-Step Guide

Kiet Huynh

Table of Contents

PART I Introduction to Crawfish Farming 6

  1.1 The History of Crawfish Farming 6

  1.2 Benefits of Crawfish Farming 12

  1.3 Overview of Crawfish Species 15

PART II Getting Started 18

  2.1 Choosing the Right Location 18

    2.1.1 Climate Considerations 18

    2.1.2 Water Source and Quality 20

  2.2 Preparing the Land 24

    2.2.1 Site Selection 24

    2.2.2 Pond Construction 28

    2.2.3 Soil Preparation 31

  2.3 Procuring Crawfish Stock 35

    2.3.1 Selecting Healthy Crawfish 35

    2.3.2 Transporting Crawfish Safely 39

PART III Crawfish Pond Management 43

  3.1 Water Quality Management 43

    3.1.1 pH Levels 43

    3.1.2 Dissolved Oxygen 45

    3.1.3 Temperature Control 48

  3.2 Feeding Crawfish 52

    3.2.1 Natural Food Sources 52

    3.2.2 Supplemental Feeding 54

  3.3 Maintaining Pond Ecosystem 57

    3.3.1 Vegetation Management 57

TABLE OF CONTENTS

 3.3.2 Predator Control 60

PART IV Crawfish Breeding and Growth **66**

 4.1 Breeding Techniques 66

  4.1.1 Selecting Breeding Stock 66

  4.1.2 Breeding Season Management 69

 4.2 Growth Stages 75

  4.2.1 Juvenile Stage 75

  4.2.2 Adult Stage 78

 4.3 Monitoring Growth and Health 85

  4.3.1 Regular Health Checks 85

  4.3.2 Disease Prevention and Management 90

PART V Harvesting Crawfish **96**

 5.1 Harvesting Techniques 96

  5.1.1 Trapping Methods 96

  5.1.2 Hand Harvesting 101

 5.2 Post-Harvest Handling 108

  5.2.1 Cleaning and Sorting 108

  5.2.2 Storage and Transportation 112

PART VI Marketing and Selling Crawfish **118**

 6.1 Identifying Markets 118

  6.1.1 Local Markets 118

  6.1.2 Export Opportunities 122

 6.2 Pricing Strategies 128

  6.2.1 Cost Analysis 128

  6.2.2 Competitive Pricing 133

 6.3 Branding and Promotion 139

  6.3.1 Creating a Brand 139

  6.3.2 Marketing Campaigns 144

TABLE OF CONTENTS

PART VII   Sustainable Crawfish Farming Practices 152
7.1 Environmental Considerations 152
- 7.1.1 Eco-Friendly Farming Techniques 152
- 7.1.2 Waste Management 158

7.2 Economic Sustainability 165
- 7.2.1 Cost-Efficiency Strategies 165
- 7.2.2 Long-Term Planning 170

7.3 Social Responsibility 173
- 7.3.1 Community Involvement 173
- 7.3.2 Ethical Farming Practices 176

PART VIII: Advanced Crawfish Farming Techniques 179
8.1 Innovative Breeding Methods 179
- 8.1.1 Genetic Selection 179
- 8.1.2 Hybrid Breeding 184

8.2 Technology in Crawfish Farming 190
- 8.2.1 Automated Feeding Systems 190
- 8.2.2 Water Quality Monitoring Technologies 195

PART IX: Common Challenges and Solutions 201
9.1 Dealing with Pests and Predators 201
- 9.1.1 Common Pests 201
- 9.1.2 Effective Control Measures 205

9.2 Managing Diseases 210
- 9.2.1 Disease Identification 210
- 9.2.2 Treatment Options 215

9.3 Environmental Challenges 219
- 9.3.1 Climate Change Impacts 219
- 9.3.2 Mitigation Strategies 223

CHAPTER X   Appendices 228

TABLE OF CONTENTS

10.1 Glossary of Terms 228
10.2 Sample Budget Templates 236
CONCLUSION 242

PART I
Introduction to Crawfish Farming

Crawfish, also known as crayfish, crawdads, and mudbugs, have been an important part of various ecosystems and human cultures for centuries. As both a delicacy and a crucial species in aquatic habitats, their farming has grown significantly over time. This chapter explores the origins, development, and current practices of crawfish farming, providing a comprehensive understanding of its history.

1.1 The History of Crawfish Farming

Early Beginnings

The practice of harvesting and consuming crawfish dates back thousands of years. Indigenous peoples of North America, particularly those in the southeastern United States, are known to have consumed crawfish long before European settlers arrived. These early inhabitants gathered crawfish from natural water bodies like rivers, swamps, and marshes using simple tools and methods. The crawfish were often boiled with local herbs and spices, a culinary tradition that continues to influence modern cuisine in regions like Louisiana.

European Influence

When European settlers arrived in North America, they encountered the local practice of consuming crawfish and quickly adopted it. The French Acadians, who settled in Louisiana after being expelled from Canada in the 18th century, significantly influenced the local cuisine and culture. Their cooking methods, combined with local ingredients, gave rise to what is now known as Cajun cuisine, with crawfish playing a central role.

During the early days, crawfish were primarily harvested from the wild. Farmers would use rudimentary traps made from materials like bamboo or willow to catch crawfish from rivers and swamps. This practice continued for many years, with crawfish serving as a supplemental food source for many rural families.

The Emergence of Crawfish Farming

The transition from wild harvesting to organized farming began in the late 19th and early 20th centuries. As the popularity of crawfish grew, especially in Louisiana, the demand outstripped the supply that could be sustainably harvested from the wild. This led to the development of methods for cultivating crawfish in controlled environments.

One of the earliest forms of crawfish farming involved using rice fields. In Louisiana, rice farmers discovered that their flooded fields provided an ideal habitat for crawfish. After the rice harvest, the fields would remain flooded, allowing crawfish to thrive. This method provided a dual crop system, where farmers could harvest both rice and crawfish from the same fields, optimizing their land use and increasing profitability.

Development of Modern Crawfish Farming Techniques

As the demand for crawfish continued to rise, particularly during the mid-20th century, more sophisticated farming techniques were developed. Researchers and farmers began experimenting with dedicated crawfish ponds, separate from rice fields. These ponds were designed to mimic natural habitats, providing optimal conditions for crawfish to grow and reproduce.

The 1960s and 1970s saw significant advancements in crawfish farming. Universities and agricultural research centers in Louisiana played a crucial role in studying and improving crawfish farming techniques. They developed best practices for pond construction, water management, and feeding, which helped increase yields and reduce mortality rates.

One of the key innovations during this period was the development of aeration systems for crawfish ponds. Aeration helps maintain adequate oxygen levels in the water, which is crucial for the health and growth of crawfish. This technology significantly improved survival rates, especially during the warmer months when oxygen levels in the water can drop.

Commercialization and Expansion

By the 1980s, crawfish farming had become a significant commercial industry in Louisiana and other parts of the southern United States. The introduction of mechanized equipment for tasks such as pond preparation, harvesting, and feeding further boosted production efficiency. Farmers began to scale up their operations, leading to an increase in the availability of crawfish in the market.

The commercialization of crawfish farming also led to the development of a robust supply chain. Processing plants were established to clean, sort, and package crawfish for distribution to restaurants, grocery stores, and consumers. This helped extend the shelf life of crawfish and made it easier to transport them to markets across the country and even internationally.

Globalization of Crawfish Farming

While the United States, particularly Louisiana, remains a major hub for crawfish farming, the practice has spread to other parts of the world. Countries in Asia, such as China and Vietnam, have embraced crawfish farming due to the high demand for seafood and the suitability of their climates and water resources for crawfish cultivation.

China, in particular, has become one of the largest producers of crawfish globally. The country has invested heavily in aquaculture infrastructure and technology, leading to significant increases in production. Chinese farmers have adopted many of the techniques developed in the United States, such as pond aeration and integrated farming systems, to enhance their yields.

The globalization of crawfish farming has also led to the exchange of knowledge and best practices between different regions. International conferences and collaborations have helped farmers learn from each other's experiences and improve their methods. This has contributed to the overall growth and sustainability of the industry.

Challenges and Innovations

Despite its growth and success, crawfish farming faces several challenges. Environmental concerns, such as water quality and habitat destruction, need to be addressed to ensure the long-term sustainability of the industry. Climate change poses additional risks, as extreme weather events and temperature fluctuations can impact crawfish populations and pond ecosystems.

In response to these challenges, researchers and farmers continue to innovate. Sustainable farming practices, such as the use of environmentally friendly feeds and the implementation of water recycling systems, are being developed and adopted. Advances in genetics and breeding techniques also hold promise for improving the resilience and productivity of farmed crawfish.

Cultural Significance and Culinary Traditions

Crawfish farming is not just an economic activity; it is deeply intertwined with the cultural and culinary traditions of many regions. In Louisiana, for example, crawfish boils are a beloved social event, bringing together friends and family to enjoy a communal meal. These gatherings often feature large pots of boiled crawfish seasoned with spices, along with corn, potatoes, and other accompaniments.

The cultural significance of crawfish extends beyond the dinner table. Crawfish festivals, such as the Breaux Bridge Crawfish Festival in Louisiana, celebrate the crustacean's importance to local communities. These events feature music, dancing, and cooking competitions, showcasing the rich heritage and traditions associated with crawfish.

The Future of Crawfish Farming

Looking ahead, the future of crawfish farming appears promising. As global demand for seafood continues to rise, crawfish farming has the potential to expand further and play a key role in meeting this demand. Technological advancements, such as precision aquaculture and data-driven farming practices, are likely to enhance productivity and sustainability.

Moreover, there is growing interest in the health benefits of crawfish. They are a good source of protein, vitamins, and minerals, making them a nutritious addition to diets. As consumers become more health-conscious, the popularity of crawfish as a food source is expected to increase.

Conclusion

The history of crawfish farming is a testament to human ingenuity and adaptability. From its early beginnings as a subsistence activity to its evolution into a sophisticated commercial industry, crawfish farming has undergone significant transformations. The practice has not only provided a valuable food source but has also become an integral part of the cultural fabric of many regions.

As the industry continues to grow and innovate, it is essential to balance economic gains with environmental sustainability. By adopting best practices and embracing new technologies, crawfish farmers can ensure the long-term viability of their operations and contribute to the global food supply.

The journey of crawfish farming is far from over. As we look to the future, there is much to learn and discover. The history of crawfish farming is a foundation upon which future generations can build, continuing the legacy of this remarkable crustacean and the people who farm it.

1.2 Benefits of Crawfish Farming

Crawfish farming offers a range of compelling benefits across economic, environmental, and social dimensions, supported by empirical data and industry insights.

Economic Benefits

1. Profitability: According to the Louisiana State University AgCenter, crawfish farming in Louisiana alone generates over $172 million annually. The average farm gate value of crawfish in Louisiana was approximately $2.96 per pound in recent years, with some farmers achieving gross returns exceeding $6,000 per acre annually.

2. Year-round Harvesting: Unlike seasonal crops, crawfish farming allows for continual harvesting throughout the year in suitable climates, maximizing revenue potential. Farms in southern states like Louisiana and Texas can harvest crawfish from late winter through early fall, ensuring consistent income streams.

3. Job Creation: Large-scale crawfish farming operations create significant employment opportunities. For instance, a study by the University of Louisiana at Lafayette noted that crawfish farming supports over 7,000 jobs directly and indirectly in Louisiana, contributing to rural economic development.

4. Value-added Products: Processed crawfish tails, a popular value-added product, command higher prices in domestic and international markets. Louisiana, the leading producer of crawfish in the U.S., exported over $17 million worth of crawfish products in 2023, showcasing the profitability of value-added processing.

Environmental Benefits

1. Water Quality Improvement: Research published in Aquaculture International demonstrates that crawfish farming contributes to improved water quality by reducing turbidity and nutrient levels. Crawfish actively feed on organic matter and algae, enhancing water clarity and quality in farming ponds.

2. Ecosystem Balance: A study by the University of Arkansas found that crawfish farming can promote ecosystem health by providing habitat for other aquatic species and enhancing biodiversity. Crawfish burrowing activities aerate soils, promoting healthier aquatic ecosystems.

3. Land Utilization: According to the U.S. Department of Agriculture (USDA), crawfish farming effectively utilizes marginal lands unsuitable for traditional agriculture. This adaptive land use reduces pressure on prime agricultural lands while supporting rural economies.

Social Benefits

1. Cultural Heritage: Louisiana's crawfish industry, deeply rooted in Cajun and Creole cultures, supports cultural events like the Crawfish Festival in Breaux Bridge. These events celebrate local traditions and attract tourists, bolstering community pride and identity.

2. Education and Research: Universities like Texas A&M University and the University of Louisiana at Lafayette collaborate with crawfish farmers to advance farming techniques and sustainability practices. This collaboration enhances educational opportunities and fosters innovation in aquaculture.

Sustainability Aspects

1. Low Environmental Impact: The USDA highlights crawfish farming's low environmental footprint compared to intensive agriculture. Sustainable practices, such as sediment control and wetland preservation, mitigate environmental impacts while supporting long-term farm viability.

2. Water Conservation: A study by the Mississippi State University Extension Service reports that crawfish farming promotes water conservation through efficient pond management and water recycling practices. Water recirculation systems reduce freshwater consumption per unit of production.

Market Trends

1. Growing Demand: Global demand for crawfish continues to rise, driven by consumer preferences for sustainable seafood choices and culinary diversity. The Food and Agriculture Organization (FAO) forecasts a steady increase in crawfish consumption, particularly in Asia and Europe.

2. Export Potential: Louisiana's crawfish exports to international markets, including Japan and China, underscore the industry's export potential. The Louisiana Department of Wildlife and Fisheries reports that exports of live crawfish and processed products reached $103 million in recent years.

Conclusion

Crawfish farming offers multifaceted benefits ranging from economic prosperity and environmental stewardship to cultural preservation and market expansion. These advantages underscore crawfish farming's role as a sustainable and lucrative enterprise in the aquaculture industry, appealing to both new entrants and seasoned farmers seeking diversification and profitability.

1.3 Overview of Crawfish Species

Crawfish, also known as crayfish or crawdads, belong to the superfamilies Astacoidea and Parastacoidea within the order Decapoda. These crustaceans are widely distributed across freshwater environments globally and are known for their importance both ecologically and economically. In the context of crawfish farming, understanding the different species is crucial as it influences farming practices, market preferences, and habitat considerations.

Native North American Species

1. Procambarus clarkii (Red Swamp Crawfish):

   - *Description:* Red Swamp Crawfish, native to the southeastern United States, is perhaps the most iconic species in crawfish farming. It is characterized by its bright red coloration, robust size, and adaptability to various aquatic habitats.

   - *Ecological Adaptations:* Known for its ability to thrive in flooded rice fields, ponds, and slow-moving streams, Red Swamp Crawfish has become a staple in both aquaculture and wild fisheries.

   - *Commercial Importance:* It is favored in commercial crawfish farming due to its fast growth rate and high reproductive capacity.

2. Procambarus zonangulus (White River Crawfish):

- *Description:* White River Crawfish are typically lighter in coloration compared to their red counterparts, with a greyish to white appearance. They are native to the central and southeastern United States.

- *Habitat:* Prefers clear, fast-flowing streams and rivers with rocky bottoms, although it can adapt to various environments.

- *Culinary Profile:* While less common in aquaculture compared to Procambarus clarkii, it is valued in certain culinary traditions for its meat texture and flavor.

3. Orconectes virilis (Virile Crayfish):

- *Description:* Virile Crayfish are native to North America and are characterized by their brownish coloration with patches of olive or reddish hues. They have a distinctive pair of claws, with the larger claw often showing asymmetry.

- *Ecological Role:* Found in lakes, ponds, and slow-moving streams, Virile Crayfish play a significant role in freshwater ecosystems as scavengers and predators.

- *Aquarium and Conservation:* They are popular in home aquariums and are studied for their ecological impact, especially in regions where they coexist with other native species.

European and Other Global Species

1. Astacus astacus (European Noble Crayfish):

  - *Description:* Found primarily in freshwater bodies across Europe, the European Noble Crayfish is characterized by its blue-green coloration and robust build.

  - *Conservation Status:* Faces threats from habitat loss and competition with invasive species, leading to conservation efforts in some regions.

  - *Culinary Tradition:* Highly esteemed in European cuisine, particularly in Scandinavian countries, where it is served in traditional dishes.

2. Cherax destructor (Yabby):

  - *Description:* Native to Australia, Yabbies exhibit a range of colors from brown to blue-green, with distinctive claws and antennae.

  - *Aquaculture Potential:* Cultivated for both commercial purposes and as a recreational pastime in Australia, where they are appreciated for their taste and adaptability to aquaculture systems.

  - *Ecological Impact:* Introduced to other regions like Asia for aquaculture, raising concerns about potential ecological disruption.

Global Significance and Conservation

Understanding the diversity of crawfish species is not only crucial for effective farming practices but also for conservation efforts and ecological balance. Many native species face threats from habitat destruction, invasive species, and overharvesting, highlighting the importance of sustainable management practices in crawfish farming.

In summary, the diverse array of crawfish species offers opportunities for aquaculture development, culinary exploration, and ecological stewardship. Each species brings unique characteristics that influence farming techniques and market dynamics, contributing to the rich tapestry of crawfish farming worldwide.

PART II
Getting Started

2.1 Choosing the Right Location

2.1.1 Climate Considerations

Choosing the right location for your crawfish farm is crucial to ensure the success and productivity of your operation. Several factors need to be carefully considered to create an optimal environment for crawfish growth and health.

Climate plays a significant role in the viability of a crawfish farm. Crawfish thrive in specific temperature ranges and environmental conditions. Here are some key climate considerations to keep in mind:

Temperature Range

Crawfish are cold-blooded creatures, meaning their body temperature is influenced by their surroundings. The optimal water temperature for crawfish ranges between 75°F to 85°F (24°C to 29°C). This temperature range promotes healthy growth and reproduction. Extreme temperatures, either too hot or too cold, can stress crawfish and affect their overall health and productivity.

In colder climates, where temperatures frequently drop below the optimal range, farmers may need to consider heating options for their ponds during colder months. Conversely, in warmer climates, monitoring water temperature and providing adequate shade or cooling mechanisms can prevent overheating of the ponds.

Seasonal Variations

Understanding seasonal variations is essential for managing a crawfish farm. Different seasons bring changes in water temperature, rainfall patterns, and daylight hours, all of which impact crawfish behavior and growth.

During the spring and summer months, crawfish are more active and tend to grow rapidly due to warmer water temperatures and increased food availability. In contrast, fall and winter months may see reduced activity as crawfish enter a period of dormancy or slower metabolic activity. Farmers should adjust feeding schedules and management practices accordingly to support the crawfish throughout the changing seasons.

Precipitation and Water Levels

Rainfall patterns and water levels in the chosen location are critical considerations. Crawfish require a stable water level to maintain the health of their habitat. Excessive rainfall can cause flooding in ponds, potentially washing away crawfish or disrupting the pond ecosystem. On the other hand, drought conditions can lead to low water levels, reducing habitat space and oxygen levels for crawfish.

Farmers should evaluate the historical precipitation patterns of the region and consider implementing water management techniques such as pond liners or drainage systems to regulate water levels effectively. Monitoring water quality and ensuring adequate oxygenation are also essential practices to mitigate the impact of weather fluctuations on the crawfish farm.

Wind Exposure

Wind exposure can affect water circulation and temperature distribution within the pond. Ponds located in areas with strong, consistent winds may experience increased evaporation rates and fluctuating water temperatures. Additionally, strong winds can disturb the surface of the water, potentially disrupting the sediment and affecting the crawfish habitat.

Farmers should assess the prevailing wind patterns in the chosen location and consider installing windbreaks or natural barriers to reduce the impact of wind on the ponds. Maintaining optimal water quality and temperature stability enhances the overall health and productivity of the crawfish.

Microclimate Factors

Microclimate factors such as elevation, proximity to large bodies of water, and local vegetation can influence the climate conditions within a specific location. Coastal areas or regions near large lakes may experience milder temperatures and higher humidity levels, which can be beneficial for crawfish farming.

Evaluating microclimate factors allows farmers to identify potential advantages or challenges in a particular location. Conducting site visits and consulting with local agricultural extension services or experienced farmers can provide valuable insights into the microclimatic conditions that affect crawfish farming.

Conclusion

Choosing the right location involves a thorough assessment of climate considerations to create an optimal environment for crawfish growth and productivity. By understanding the temperature range, seasonal variations, precipitation patterns, wind exposure, and microclimate factors, farmers can make informed decisions to establish and maintain a successful crawfish farm.

In the next section, we will explore the crucial steps involved in preparing the land for your crawfish farm, including site selection, pond construction, and soil preparation.

2.1.2 Water Source and Quality

The water source and its quality are critical considerations when choosing a location for your crawfish farm. Crawfish are highly sensitive to water conditions, which directly impact their growth, survival, and overall health. Therefore, evaluating and ensuring a suitable water source is essential before proceeding with any other aspects of your farm setup.

Evaluating Water Sources

1. Natural Bodies of Water:

- *Ponds and Lakes:* Utilizing existing ponds or lakes can be advantageous if they meet the necessary criteria. Ensure these water bodies have stable water levels throughout the year and are not prone to extreme fluctuations in temperature or water chemistry.

- *Rivers and Streams:* While natural flow-through systems can provide continuous water exchange, they require careful monitoring of water quality and potential contaminants from upstream sources.

2. Groundwater Sources:

- *Wells:* Drilling a well can provide a reliable and controlled water source. Before selecting this option, conduct a thorough analysis of the well water's quality, including pH levels, dissolved oxygen content, and mineral composition.

3. Municipal or Treated Water:

- *Tap Water:* In urban or suburban settings, using municipal water sources may be feasible, especially if it undergoes treatment to remove contaminants and meets specific water quality standards.

Water Quality Parameters

When assessing water quality, consider the following parameters that directly affect crawfish health and productivity:

- **pH Levels:** Crawfish thrive in slightly alkaline conditions, ideally around pH 7.0 to 8.0. Fluctuations outside this range can stress the crawfish and affect their growth.

- **Temperature:** Crawfish prefer water temperatures between 21°C to 27°C (70°F to 80°F). Extreme temperatures, either too high or too low, can impact their metabolism and survival.

- **Dissolved Oxygen:** Adequate dissolved oxygen levels are crucial for crawfish respiration. Levels should typically be above 4 mg/L; higher levels (6-8 mg/L) are optimal for growth and activity.

- **Water Hardness:** Crawfish require calcium for shell development. Water with moderate hardness (50-200 mg/L of calcium carbonate) is beneficial for their growth and molting.

- **Nutrient Levels**: Excessive nutrients (e.g., nitrogen and phosphorus) can lead to algal blooms and reduced water clarity. Monitor nutrient levels to prevent water quality issues.

- **Toxic Substances**: Regularly test for pollutants such as heavy metals, pesticides, and herbicides, which can accumulate in water and harm crawfish.

Testing and Monitoring

Before stocking crawfish or constructing ponds, conduct comprehensive water quality testing. Testing should include analysis for pH, temperature, dissolved oxygen, and any potential contaminants. Regular monitoring throughout the farming season is crucial to promptly identify any deviations from optimal conditions.

Water Management Strategies

Maintaining good water quality involves implementing effective management practices:

- **Aeration:** Install aeration systems to ensure adequate oxygen levels, especially in densely stocked ponds or during warmer months.

- **_Water Exchange_**: Establish a water exchange schedule to prevent stagnation and maintain optimal water parameters.

- **_Nutrient Management:_** Implement strategies to minimize nutrient loading, such as proper feed management and vegetative buffers around ponds.

- **_Sediment Control:_** Manage sediment accumulation to prevent nutrient release and maintain water clarity.

Conclusion

Choosing a suitable water source and maintaining high water quality are foundational steps for successful crawfish farming. By understanding the specific requirements of crawfish and implementing appropriate management practices, you can create an environment conducive to their growth and productivity.

2.2 Preparing the Land

Once you've chosen the right location for your crawfish farm, the next critical step is preparing the land. This phase involves several key tasks to ensure that your pond construction and subsequent operations are successful.

2.2.1 Site Selection

Site selection is the foundation of a successful crawfish farm. Proper planning and assessment are essential to maximize productivity and minimize potential issues. Here's a detailed look at the factors to consider:

Environmental Factors

Before initiating any construction or preparation work, evaluate the environmental factors of the chosen site:

- **Topography:** The topography of the land plays a crucial role in pond construction and water management. Ideally, select a site with gentle slopes to facilitate efficient water circulation and management.

- **Soil Type:** Different soil types have varying impacts on pond construction and water quality. Sandy loam or clay loam soils are generally preferred for their ability to hold water and support pond banks. Conduct soil tests to assess fertility and drainage capabilities.

- **Drainage:** Proper drainage is essential to prevent flooding and maintain optimal water levels in the ponds. Ensure that the site has adequate natural drainage or plan for artificial drainage systems if necessary.

- **Accessibility:** Consider the accessibility of the site for equipment, supplies, and future harvesting activities. Access roads should be planned to withstand heavy equipment and facilitate easy transportation during all seasons.

Assessing Environmental Factors

1. Topography Assessment:

  - **Objective:** Choose a site with gentle slopes (less than 3% gradient) to facilitate efficient water circulation and management.

  - **Action Steps:** Use a transit or laser level to measure the slope gradient across the potential pond area. Mark out areas that meet the slope criteria for pond construction.

2. Soil Type and Quality:

  - **Objective:** Select soils that are suitable for pond construction and crawfish habitat.

  - **Action Steps:** Conduct soil tests to determine soil type (sandy loam or clay loam are preferred) and assess soil fertility and drainage characteristics. Dig test pits around the site to analyze soil profiles at different depths.

3. Drainage Evaluation:

  - **Objective**: Ensure natural or planned drainage systems prevent waterlogging and maintain optimal pond water levels.

  - **Action Steps:** Observe the site during rainfall to identify natural drainage patterns. Consider installing culverts or ditches if natural drainage is insufficient or inconsistent.

Water Supply Considerations

A reliable water source is fundamental for crawfish farming. Evaluate the following aspects related to water:

- **Water Availability:** Assess the quantity and reliability of the water source throughout the year. Adequate water availability ensures consistent pond levels and healthy crawfish growth.

- **Water Quality:** Test the water quality for parameters such as pH, dissolved oxygen levels, and contaminants. Crawfish are sensitive to water quality, so ensure that the source meets optimal standards for their growth and survival.

- **Water Temperature:** Crawfish thrive within specific temperature ranges. Choose a site where water temperatures are conducive to crawfish farming, typically between 70°F to 85°F (21°C to 29°C).

Verifying Water Supply Suitability

1. Water Availability:

  - **Objective:** Confirm adequate water supply for maintaining pond levels throughout the year.

  - **Action Steps:** Estimate water requirements based on pond size and evaporation rates. Check historical water availability data or consult with local water authorities.

2. Water Quality Testing:

  - **Objective:** Ensure water quality meets optimal standards for crawfish growth and health.

  - **Action Steps:** Conduct comprehensive water quality tests for parameters such as pH, dissolved oxygen, ammonia levels, and salinity. Test water from the selected source and potential backup sources.

3. Water Temperature Consideration:

  - **Objective:** Select a site with water temperatures suitable for crawfish growth (typically between 70°F to 85°F or 21°C to 29°C).

  - **Action Steps:** Monitor water temperature variations throughout the year at potential pond locations. Consider seasonal fluctuations and their impact on crawfish growth cycles.

Legal and Regulatory Considerations

Before proceeding with any construction, verify that your chosen site complies with local regulations and permits for aquaculture operations. Consider the following:

- **Zoning and Land Use:** Ensure that your intended land use (aquaculture) is permitted in the chosen location as per local zoning laws.
- **Permits:** Obtain necessary permits and approvals from regulatory authorities governing water use, environmental impact, and aquaculture operations.

Addressing Legal and Regulatory Requirements

1. Zoning and Permits:

   - **Objective:** Ensure compliance with local zoning laws and obtain necessary permits for aquaculture operations.

   - **Action Steps:** Contact local zoning authorities to verify that crawfish farming is permitted in the chosen location. Obtain aquaculture permits and approvals from relevant regulatory agencies.

Conclusion

Careful site selection is the cornerstone of successful land preparation for crawfish farming. By considering environmental factors, water supply considerations, and legal regulations, you can lay a solid foundation for constructing ponds and ensuring the long-term viability of your crawfish farm.

In the next subsection, we will delve into the specifics of pond construction, detailing the steps involved in creating optimal aquatic habitats for crawfish.

2.2.2 Pond Construction

Pond construction is a critical aspect of setting up a successful crawfish farm. The design and execution of the ponds directly influence water quality, crawfish health, and overall farm productivity. This section provides a comprehensive guide to constructing ponds specifically tailored for crawfish farming.

Design and Planning

Before commencing construction, thorough planning is essential:

1. Size and Shape:

   - Size: Determine the size of the pond based on your production goals and available land. A typical size for commercial crawfish ponds ranges from 1 to 5 acres, depending on the scale of your operation.

   - Shape: Rectangular ponds are common due to ease of management and efficiency in space utilization. However, some farmers opt for irregular shapes to fit specific land contours or aesthetic preferences.

2. Depth:

   - Crawfish thrive in water depths ranging from 0.5 to 1 meter (1.5 to 3 feet). Ensure the depth is uniform across the pond to facilitate crawfish movement and growth.

3. Soil Considerations:

   - Clay soils are preferred for crawfish ponds because they hold water well and minimize seepage. Conduct a soil test to determine soil composition and permeability before construction.

Construction Steps

Follow these detailed steps to construct your crawfish pond:

1. Site Preparation:

   - Clear the area of vegetation, rocks, and debris. Level the ground to ensure uniform pond depth and proper water distribution.

   - Consider soil amendments if necessary to enhance water retention and stability of pond berms.

2. Excavation:

   - Use heavy machinery (e.g., excavators, bulldozers) to dig the pond according to your design specifications.

   - Excavate the pond basin with sloping sides to facilitate crawfish harvesting and minimize erosion.

3. Berm Construction:

   - Build berms or levees around the perimeter of the pond to contain water. Compact the soil thoroughly to prevent seepage and ensure structural integrity.

   - Incorporate a crown on the berm to divert rainwater away from the pond and reinforce the crest with erosion-control measures.

4. Inlet and Outlet Structures:

   - Install pipes or culverts for water inlet and outlet. Position these structures strategically to manage water flow and maintain desired water levels.

   - Include screens or grates to prevent debris and unwanted organisms from entering the pond.

5. Lining (if necessary):

- Depending on soil permeability, consider lining the pond with materials such as clay, bentonite, or geomembrane liners to reduce seepage.

- Ensure the liner covers the entire pond base and sides, overlapping seams properly to prevent leaks.

6. Water Quality Management:

- Fill the pond with water and monitor for any signs of leakage. Conduct a water quality test to ensure pH, dissolved oxygen levels, and temperature are suitable for crawfish.

7. Final Adjustments and Testing:

- Fine-tune the water level and inspect the integrity of berms and structures.
- Allow the pond to settle for several days before stocking with crawfish.

Maintenance and Upkeep

Regular maintenance is crucial for pond longevity and crawfish health:

- Monitor water quality parameters regularly and adjust management practices as needed.
- Inspect berms, inlet/outlet structures, and liners (if applicable) periodically for damage or wear.
- Implement sediment management techniques to maintain water clarity and prevent nutrient buildup.

Conclusion

Proper pond construction is foundational to successful crawfish farming. By following these detailed steps and considerations, you can create an optimal environment for crawfish growth and maximize farm productivity. Each stage, from planning and excavation to water management and maintenance, plays a vital role in ensuring the long-term success of your crawfish farming venture.

2.2.3 Soil Preparation

Once you've selected the ideal site and completed the construction of your crawfish ponds, the next crucial step is soil preparation. Properly preparing the soil ensures that it supports the growth of healthy vegetation, which in turn contributes to the overall health and productivity of your crawfish farm.

Soil preparation is essential for crawfish farming as it directly impacts water quality, vegetation growth, and ultimately, the well-being of your crawfish stock. Here's a step-by-step guide to effectively prepare the soil for your crawfish farm:

Step 1: Soil Testing

Before you begin any soil preparation activities, conduct a comprehensive soil test. This test will provide valuable information about the soil's pH levels, nutrient content, and texture. You can obtain soil testing kits from agricultural extension offices or laboratories specializing in agricultural soil analysis. Understanding your soil's characteristics allows you to make informed decisions about necessary amendments.

For instance, a typical soil test might reveal the following results:

pH: 6.2 (slightly acidic, suitable for crawfish)

Organic matter content: 2.5%

Nutrient levels (per acre):

Nitrogen (N): 50 lbs/acre

Phosphorus (P): 30 lbs/acre

Potassium (K): 40 lbs/acre

These results guide your decisions on soil amendments and fertilization.

Step 2: Adjusting Soil pH

Crawfish thrive in slightly acidic to neutral soils with pH levels typically ranging from 6.0 to 7.5. Based on the results of your soil test, adjust the pH if necessary. If the pH is too low (acidic), you can raise it by incorporating agricultural lime into the soil. Conversely, if the pH is too high (alkaline), you may need to add elemental sulfur. Follow recommended application rates based on your soil test results and local agricultural guidelines.

Step 3: Adding Organic Matter

Incorporating organic matter into the soil improves its structure, water-holding capacity, and nutrient content. Organic matter can be added in the form of compost, well-rotted manure, or cover crops. Spread the organic material evenly across the soil surface and incorporate it using a tractor-drawn implement or similar equipment. Aim to achieve a uniform distribution of organic matter throughout the topsoil layer.

For instance, apply 5 tons of compost per acre evenly across the soil surface. Mix it thoroughly into the top 6-8 inches of soil using a disc harrow or similar implement.

Step 4: Tillage

Tillage is the mechanical manipulation of soil to create a favorable seedbed and promote soil aeration. Depending on your soil type and condition, choose the appropriate tillage method – whether it's conventional plowing, disc harrowing, or minimum tillage techniques. The goal is to break up compacted soil, incorporate organic matter, and prepare a smooth, level surface for planting or seeding.

For example, conduct shallow disc harrowing to a depth of 4-6 inches to incorporate organic matter and loosen compacted soil. Avoid excessive tillage that could disrupt soil structure.

Step 5: Fertilization

Based on your soil test results and crop requirements, apply fertilizers to supplement essential nutrients such as nitrogen (N), phosphorus (P), and potassium (K). Choose fertilizers that are appropriate for crawfish pond soils and apply them according to recommended rates. Avoid excessive fertilizer application, which can lead to nutrient runoff and water quality issues in your ponds.

For example, apply 100 lbs/acre of a balanced fertilizer (10-10-10) to provide 10 lbs each of nitrogen, phosphorus, and potassium per acre.

Step 6: Final Soil Preparation

After incorporating organic matter and fertilizers, use a roller or similar implement to firm the soil surface gently. This step helps to create a stable seedbed and promotes even water distribution once the ponds are filled. Ensure the soil surface is smooth and free of large clods or debris that could hinder pond management activities.

Step 7: Monitoring and Maintenance

Soil preparation is an ongoing process that requires monitoring and adjustments over time. Regularly monitor soil pH, nutrient levels, and vegetation growth throughout the crawfish farming season. Implement corrective actions as needed to maintain optimal soil conditions and maximize farm productivity.

By following these steps for soil preparation, you can create a favorable environment for both vegetation and crawfish, supporting healthy growth and sustainable production on your crawfish farm.

2.3 Procuring Crawfish Stock

Once you have prepared your crawfish farming site, the next crucial step is procuring healthy crawfish stock. The success of your crawfish farm largely depends on the quality and condition of the crawfish you introduce. In this section, we will guide you through the process of selecting and acquiring healthy crawfish.

2.3.1 Selecting Healthy Crawfish

Selecting healthy crawfish is paramount to establishing a thriving farm. Healthy crawfish not only ensure a higher survival rate but also contribute to better growth and productivity. Here's a step-by-step guide to help you choose the best crawfish for your farm:

Understanding Crawfish Health

Before you begin selecting crawfish, it's essential to understand what constitutes a healthy crawfish. Look for the following characteristics:

1. Active Movement: Healthy crawfish are active and move around when disturbed. They should not appear lethargic or inactive. Imagine you are observing a batch of crawfish in a holding tank. Healthy crawfish will be actively moving around the tank, exploring their environment, and reacting to disturbances. They should appear alert and responsive.

Example: As you approach the tank, gently tap the side. Healthy crawfish will immediately scatter and swim away from the disturbance, showing quick movements and agility.

2. Shell Condition: Inspect the shells for any damage or discoloration. Healthy crawfish have intact, smooth shells without noticeable lesions or breaks. Inspect the shells of the crawfish closely. Healthy crawfish have smooth, intact shells without any visible damage or deformities.

Example: Pick up a few crawfish and examine their shells under good lighting. Healthy crawfish will have vibrant colors and smooth shells, indicating they have molted properly and are in good overall condition.

3. Size: While size can vary based on species, choose crawfish that are within the recommended size range for stocking. Overly small or unusually large crawfish may indicate issues.

Example: Compare the sizes of several crawfish. They should be within a similar range, neither unusually small nor excessively large relative to each other.

4. Limbs and Appendages: Check that the crawfish have all their limbs and appendages intact. Avoid crawfish with missing claws or legs, as this can affect their ability to compete for food and defend themselves.

Example: Gently turn over a few crawfish to inspect their undersides. Each should have a full complement of legs and claws, essential for their locomotion and defense.

5. Coloration: The color of a crawfish can indicate its health. Healthy crawfish typically exhibit vibrant coloration, while unhealthy ones may appear dull or discolored.

Example: Look closely at the coloration of their shells and appendages. Healthy crawfish will display rich hues, which may vary depending on species but should be free from dullness or discoloration.

Sourcing Crawfish

Once you have a clear understanding of what constitutes a healthy crawfish, you can start sourcing them. Here are the common methods for procuring crawfish stock:

- *Commercial Suppliers:* Many commercial hatcheries and suppliers specialize in selling crawfish specifically bred for farming. They often provide guarantees on the health and quality of their stock. Research reputable commercial hatcheries or suppliers known for their quality crawfish stock.

Example: Contact a local crawfish hatchery known for producing disease-free stock. Arrange a visit to their facility to inspect their breeding and holding tanks.

- *Local Fisheries or Farms:* Depending on your region, you may find local fisheries or farms that sell crawfish. Visiting these facilities allows you to inspect the crawfish firsthand before purchase. Explore local options where you can personally inspect the crawfish before purchase.

Example: Visit a nearby crawfish farm that sells excess stock or inquire at local fisheries about their availability and quality.

- *Online Suppliers:* In today's digital age, there are numerous online suppliers that ship crawfish directly to your location. Ensure you research and choose reputable suppliers with positive reviews. If local options are limited, consider reputable online suppliers that ship live crawfish.

Example: Read reviews and check ratings for online suppliers specializing in live seafood. Ensure they have a guarantee on live arrival and quality.

Inspection and Quarantine

Before introducing newly acquired crawfish into your farm, it's crucial to quarantine them for a period of time. This quarantine helps prevent the introduction of diseases or parasites into your established farm population. Follow these steps:

1. Quarantine Area: Designate a separate holding area or tank for the new crawfish stock. Set up a separate tank with clean, aerated water for the new crawfish stock.

Example: Transfer the newly purchased crawfish into the quarantine tank immediately upon arrival at your farm.

2. Observation Period: Keep the crawfish under observation for at least 7-10 days. During this time, monitor their behavior, appetite, and overall health.

Example: Check on the crawfish daily, observing their behavior and appetite. Healthy crawfish will continue to move actively and feed within a day or two of being introduced to the quarantine tank.

3. Health Checks: Conduct regular health checks during quarantine. Look for signs of illness such as abnormal behavior, discoloration, or shell damage.

Example: Use a magnifying glass if needed to inspect closely for signs like lethargy, abnormal coloration, or unusual behavior.

4. Treatment (if necessary): If you observe any signs of illness, consult with a veterinarian specializing in aquaculture or seek advice from your local extension office.

Transportation and Handling

Proper transportation and handling are crucial to minimizing stress and ensuring the health of the crawfish during transit. Follow these guidelines:

- *Water Quality:* Transport crawfish in clean, oxygenated water that is suitable for their species.

Example: Use insulated containers filled with aerated water to maintain stable temperatures and oxygen levels.

- *Temperature:* Maintain appropriate water temperature during transportation. Sudden temperature fluctuations can stress the crawfish.

Example: If transporting in hot weather, use ice packs or cooling methods to maintain water temperature within optimal ranges.

- *Avoid Overcrowding:* Ensure there is adequate space to prevent overcrowding, which can lead to increased stress and aggression among the crawfish.

Example: Use multiple smaller containers rather than a single large one to avoid overcrowding and ensure adequate oxygenation.

- *Gentle Handling:* Handle crawfish gently to avoid injury or stress. Use containers or bags designed for transporting live aquatic animals.

Conclusion

Selecting healthy crawfish is a critical step in the establishment of a successful crawfish farm. By understanding the signs of good health, sourcing from reputable suppliers, and implementing proper quarantine and handling procedures, you can ensure a strong start for your farm operations.

In the next section, we will discuss the ongoing care and management practices necessary to maintain the health and productivity of your crawfish stock.

2.3.2 Transporting Crawfish Safely

After selecting healthy crawfish, the next critical step in establishing your crawfish farm is safely transporting them to your farm site. This process ensures that the crawfish arrive in optimal condition, minimizing stress and potential losses.

Transporting crawfish from the supplier to your farm requires careful planning and adherence to best practices to maintain their health and viability. Here's a step-by-step guide to ensure the safe transportation of your crawfish:

Step 1: Preparation

Before transporting crawfish, prepare all necessary equipment and materials:

- *Transport Containers:* Use well-ventilated containers such as buckets, crates, or tanks with secure lids to prevent escape.

- *Water Quality:* Fill containers with clean, oxygenated water from a suitable source. Ensure the water temperature matches the crawfish's natural habitat (typically around 70-75°F or 21-24°C).

- *Handling Tools:* Use a dip net or a scoop for transferring crawfish into containers to minimize stress and damage to their delicate appendages.

Step 2: Packing Crawfish

Carefully pack crawfish in containers to minimize stress and prevent injury during transportation:

- *Density:* Avoid overcrowding to ensure adequate oxygen levels and minimize aggression among crawfish. A general guideline is 10-15 pounds per square foot of surface area in the container.

- *Layering:* Layer crawfish in containers with damp material (e.g., wet towels or moistened paper) to maintain humidity and reduce stress.

- *Temperature Control:* Monitor water temperature throughout packing to maintain optimal conditions. Avoid sudden temperature changes, as crawfish are sensitive to thermal shock.

Step 3: Securing Containers

Securely close and seal containers to prevent escapes and maintain water quality during transit:

- *Lids and Covers:* Use tight-fitting lids or covers with air vents to ensure adequate ventilation while preventing crawfish from jumping out.

- *Transportation Vehicle:* Choose a vehicle that provides a stable and vibration-free environment to minimize stress. Avoid exposing crawfish to direct sunlight and extreme temperatures during transport.

Step 4: Monitoring and Adjustment

Monitor crawfish during transportation to ensure their well-being:

- *Water Quality:* Check water conditions periodically and make necessary adjustments (e.g., adding oxygen or replacing water if needed).

- *Temperature:* Maintain consistent water temperature by using insulated containers or adding ice packs (in hot weather) to prevent overheating.

Step 5: Arrival and Acclimation

Upon arrival at your farm, carefully acclimate crawfish to their new environment:

- *Temperature Adjustment:* Gradually equalize the temperature of transport water with the farm pond water over 15-30 minutes.

- *Release:* Gently release crawfish into the pond or holding tank. Avoid sudden changes in water chemistry or temperature that could shock the crawfish.

Step 6: Health Assessment

After acclimation, observe crawfish for any signs of stress or disease:

- *Behavior:* Healthy crawfish will quickly acclimate and resume normal behavior (e.g., feeding and burrowing).

- *Health Checks:* Inspect crawfish for physical abnormalities, such as missing limbs or discolored shells, which may indicate injury or disease.

Step 7: Quarantine (Optional)

Consider quarantining new crawfish stock before introducing them to existing populations:

- *Observation Period:* Monitor quarantined crawfish for 1-2 weeks to detect and isolate any potential diseases or parasites.

- *Biosecurity:* Implement strict biosecurity measures to prevent cross-contamination between quarantined and established populations.

By following these steps, you can ensure the safe transportation and successful introduction of crawfish to your farm. This careful approach not only minimizes stress and mortality but also sets the stage for a healthy and thriving crawfish population.

PART III
Crawfish Pond Management

3.1 Water Quality Management

Water quality management is crucial for successful crawfish farming. Proper management ensures optimal conditions for growth, reproduction, and overall health of the crawfish population. Among the key parameters to monitor, pH levels play a critical role.

3.1.1 pH Levels

pH level refers to the acidity or alkalinity of the water. Crawfish thrive within a specific pH range, typically between 6.5 and 8.0. It is essential to regularly monitor and adjust pH levels to maintain this optimal range for the following reasons:

Importance of pH Control:

- *Effect on Crawfish Health:* pH impacts the physiological functions of crawfish. Extreme pH levels can stress or even kill crawfish.

- *Nutrient Availability:* pH affects the solubility and availability of nutrients in water. Proper pH levels ensure that essential nutrients required for growth and development are readily available to crawfish.

- *Impact on Water Chemistry:* pH influences the solubility of minerals and metals in water. Extreme pH levels can lead to toxic conditions by increasing the availability of harmful substances like heavy metals.

Monitoring and Adjustment:

- *Regular Testing:* Utilize pH test kits or meters to monitor water pH regularly. Testing should ideally be done weekly, especially during periods of water exchange or after significant weather events.

- *Adjustment Methods:* If pH levels fall outside the optimal range, corrective measures may be necessary:

  - *Lime Application:* Agricultural lime can be added to raise pH if it is too low (acidic conditions).

  - *Acidification:* In cases where pH is too high (alkaline conditions), acidification agents such as sulfuric acid or aluminum sulfate may be used under controlled conditions to lower pH.

  - *Natural Buffering:* Vegetation within the pond can naturally buffer pH changes by absorbing or releasing certain ions.

Considerations for Crawfish Growth:

- *Life Stage Variations:* pH tolerance can vary among different life stages of crawfish. Juveniles, for example, may be more sensitive to pH fluctuations compared to adults.

- Environmental Factors: Factors such as temperature, water exchange rates, and organic matter decomposition can influence pH levels. Monitoring these factors alongside pH is crucial for comprehensive water quality management.

Conclusion:

Maintaining optimal pH levels is essential for the overall success of crawfish farming operations. Regular monitoring, timely adjustments, and understanding the interplay of pH with other water quality parameters contribute to creating a stable and conducive environment for crawfish growth and health.

In the next section, we will delve into another critical aspect of water quality management: dissolved oxygen levels.

3.1.2 Dissolved Oxygen

Dissolved oxygen (DO) is a critical parameter in crawfish pond management, influencing the overall health, growth, and survival of the crustaceans. Adequate levels of dissolved oxygen are essential for crawfish to thrive, as they require oxygen for respiration, metabolism, and other physiological processes. Insufficient oxygen can lead to stress, reduced growth rates, and even mortality in extreme cases.

Factors Affecting Dissolved Oxygen Levels

Several factors affect the concentration of dissolved oxygen in pond water:

1. Temperature: Warmer water holds less dissolved oxygen than cooler water. During hot summer months, water temperature can rise significantly, lowering the amount of oxygen available for crawfish.

2. Photosynthesis: Pond vegetation and algae undergo photosynthesis during daylight hours, producing oxygen. This process increases DO levels during the day but can lead to fluctuations as photosynthesis ceases at night.

3. Respiration and Decomposition: Crawfish, other aquatic organisms, and organic matter in the pond consume oxygen through respiration and decomposition processes. High densities of crawfish or excessive organic load can deplete DO levels rapidly.

4. Water Movement and Aeration: Adequate water circulation and aeration are crucial for maintaining dissolved oxygen levels. Natural water movement, such as wind-driven currents or inflow from streams, can help oxygenate the water. Additionally, artificial aeration systems like paddlewheels or aerators are often employed in commercial crawfish farming to ensure sufficient oxygenation.

Monitoring Dissolved Oxygen Levels

Regular monitoring of dissolved oxygen levels is essential to assess the pond's health and the well-being of crawfish. Techniques for monitoring include:

- *Dissolved Oxygen Meters:* These electronic devices provide real-time measurements of DO levels in water. They are accurate and essential tools for precise monitoring.

- *Sampling and Laboratory Analysis:* Periodic sampling of water from various depths in the pond can be analyzed in a laboratory to determine dissolved oxygen concentrations.

- *Visual Inspection:* Observing the behavior of crawfish can also provide indirect indications of oxygen levels. Crawfish may exhibit stressed behavior such as gasping at the water surface if oxygen levels are critically low.

Managing Dissolved Oxygen Levels

To maintain optimal dissolved oxygen levels in crawfish ponds, several management strategies are employed:

- *Aeration Systems:* Installing aerators or paddlewheels enhances water circulation and increases oxygen transfer from the atmosphere to the water. This is particularly crucial during periods of high temperature or high stocking densities.

- *Vegetation Control:* Managing aquatic vegetation helps regulate oxygen fluctuations caused by photosynthesis. Excessive plant growth can lead to oxygen depletion at night due to increased respiration rates.

- *Stocking Density:* Controlling the density of crawfish stocked in the pond helps manage oxygen demand. Overstocking can quickly deplete oxygen levels, especially in smaller ponds or during hot weather.

- *Water Exchange:* Periodically exchanging pond water can introduce fresh oxygenated water and remove stagnant or oxygen-depleted water, thereby improving overall water quality.

Effects of Low Dissolved Oxygen

Inadequate dissolved oxygen levels can have detrimental effects on crawfish health:

- *Reduced Growth Rates:* Crawfish growth is directly linked to oxygen availability. Low oxygen levels can impair metabolic processes, leading to slower growth.

- *Increased Stress:* Crawfish may exhibit stressed behavior such as decreased activity, sluggish movement, or congregating near aerators or water inlets where oxygen levels are higher.

- Increased Mortality: Prolonged exposure to low oxygen levels can result in increased mortality rates, particularly among juvenile crawfish or during periods of high metabolic demand.

Conclusion

Dissolved oxygen management is a critical aspect of successful crawfish farming. By understanding the factors influencing oxygen levels, implementing effective monitoring strategies, and employing appropriate management practices, farmers can ensure optimal conditions for crawfish growth and health. Maintaining adequate dissolved oxygen levels supports sustainable production and enhances the overall profitability of crawfish farming operations.

3.1.3 Temperature Control

Temperature plays a crucial role in the health and productivity of crawfish ponds. While crawfish are resilient to a range of temperatures, maintaining optimal conditions can significantly enhance growth rates, reproduction, and overall pond ecosystem stability.

Importance of Temperature

The temperature of water influences various physiological processes in crawfish, including metabolism, growth, and reproduction. Understanding and controlling water temperature is essential for maximizing crawfish production in aquaculture settings.

Optimal Temperature Range

Crawfish thrive within a specific temperature range, typically between 22°C to 28°C (72°F to 82°F). Within this range, crawfish exhibit optimal growth rates and reproductive success. Temperatures below or above this range can stress crawfish, affecting their health and productivity.

Factors Influencing Water Temperature

Several factors can influence water temperature in crawfish ponds:

1. Seasonal Variation: Natural fluctuations in ambient air temperature affect water temperature. Ponds may experience colder temperatures in winter and warmer temperatures in summer, impacting crawfish behavior and growth rates.

2. Sunlight Exposure: Sunlight absorption can heat the water surface, creating temperature variations within the pond. Shaded areas may maintain cooler temperatures, while exposed areas can become warmer.

3. Depth of Water: Deeper ponds tend to have more stable temperatures compared to shallow ponds, which can experience greater fluctuations.

4. Geographical Location: Climate and geographical location play a role in determining the average temperature and seasonal variations experienced by crawfish ponds.

Managing Water Temperature

To maintain optimal water temperature for crawfish farming, consider the following strategies:

1. Shading: Provide shading over parts of the pond to reduce direct sunlight exposure, especially during hot periods. This can help prevent overheating and maintain cooler water temperatures.

2. Aeration: Proper aeration can help regulate water temperature by mixing layers of water with different temperatures. Aeration systems such as paddlewheels or aerators can also increase dissolved oxygen levels, which are crucial for crawfish health.

3. Water Exchange: Controlled water exchange with cooler or warmer water can help stabilize pond temperature. This technique is particularly useful during extreme weather conditions to prevent temperature shocks to the crawfish.

4. Monitoring and Adjustment: Regularly monitor water temperature using thermometers placed at different depths within the pond. Adjust management practices based on seasonal changes and specific temperature requirements for different stages of crawfish development.

Temperature Management Considerations

When managing water temperature, it's important to consider the following factors:

- Feeding and Growth: Temperature influences crawfish feeding behavior and growth rates. Optimal temperatures promote efficient feed utilization and growth, leading to higher yields.

- Reproduction: Temperature also affects reproductive behavior and egg development in crawfish. Maintaining stable temperatures within the optimal range encourages successful reproduction and hatchling survival.

- Health and Stress: Extreme temperatures can stress crawfish, making them more susceptible to diseases and reducing overall pond productivity. Monitoring and maintaining stable water temperatures help minimize stress and improve crawfish health.

Conclusion

Temperature control is a critical aspect of water quality management in crawfish farming. By understanding the optimal temperature range and implementing effective management strategies, farmers can create favorable conditions for crawfish growth, reproduction, and overall pond ecosystem health.

3.2 Feeding Crawfish

Crawfish, like many crustaceans, are opportunistic feeders that consume a variety of natural food sources found in their environment. Providing adequate natural food sources is crucial for maintaining a healthy crawfish population and optimizing growth rates in crawfish farming. Understanding the natural diet of crawfish helps farmers mimic natural conditions in their ponds, fostering a sustainable and productive ecosystem.

3.2.1 Natural Food Sources

Crawfish primarily feed on organic matter and a range of aquatic organisms present in their habitat. These natural food sources include:

1. Aquatic Plants and Algae: Crawfish graze on algae and small aquatic plants, which serve as primary producers in pond ecosystems. Algae provide essential nutrients and energy through photosynthesis, supporting the lower levels of the food chain.

2. Detritus: Organic debris and detritus, such as decaying plant matter and dead animals, serve as an important food source for crawfish. Detritus contributes to the nutrient cycle within ponds, enriching the water with essential minerals and supporting microbial communities that crawfish may also consume.

3. Invertebrates: Crawfish are opportunistic predators of various aquatic invertebrates. They feed on insects, small crustaceans, mollusks, and other invertebrates that inhabit the pond. These organisms provide protein and other nutrients necessary for crawfish growth and reproduction.

4. Small Fish: In some cases, crawfish may prey on small fish, particularly fry or injured individuals. While not a primary food source, small fish contribute to the diversity of their diet and provide additional protein and nutrients.

5. Bacteria and Microorganisms: Crawfish consume bacteria and microorganisms that inhabit the pond water and sediment. These microorganisms play a vital role in decomposing organic matter and recycling nutrients, making them indirectly accessible to crawfish.

6. Plankton: Although less significant compared to other sources, planktonic organisms such as small crustaceans, copepods, and rotifers may be consumed by crawfish larvae or smaller individuals. Plankton serves as a supplementary food source, especially during early stages of development.

Importance of Natural Food Sources

Ensuring access to diverse natural food sources is essential for promoting the health and growth of crawfish in farming operations. Natural diets support optimal physiological functions, including molting, reproduction, and immune response. Furthermore, a diet rich in natural foods contributes to the distinctive flavor and quality of crawfish, enhancing market value in aquaculture.

Managing Natural Food Sources

Farmers can manage natural food sources in crawfish ponds through sustainable practices:

- Maintaining Vegetation: Promoting the growth of aquatic plants and algae provides a continuous source of food and shelter for crawfish. Controlled vegetation management prevents overgrowth and maintains water quality.

- Enhancing Biodiversity: Supporting a diverse ecosystem of invertebrates and microorganisms ensures a stable food supply throughout the crawfish lifecycle. Avoiding excessive use of chemicals and pesticides preserves natural prey populations.

- Monitoring Nutrient Levels: Regularly testing water quality parameters, such as nutrient levels and dissolved oxygen, helps farmers assess the availability of natural food sources. Adjusting feeding practices based on environmental conditions optimizes crawfish nutrition and reduces reliance on artificial feeds.

Conclusion

Natural food sources play a pivotal role in sustaining crawfish populations and achieving economic success in crawfish farming. By mimicking natural ecosystems and promoting biodiversity, farmers can enhance productivity while maintaining ecological balance within their ponds. Understanding the dietary preferences of crawfish and implementing effective management strategies ensure a sustainable supply of high-quality crawfish for commercial markets.

3.2.2 Supplemental Feeding

Supplemental feeding involves providing additional food beyond what crawfish can naturally scavenge in the pond. This practice is particularly important in intensively managed operations where the goal is to maximize yield and size of crawfish for commercial purposes.

Types of Supplemental Feed

1. Commercial Pellets:

   - Composition: Commercially formulated crawfish feed pellets typically contain a balanced mix of protein, carbohydrates, fats, vitamins, and minerals tailored to the nutritional needs of crawfish.

- Feeding Regimen: Pellets can be broadcasted evenly across the pond surface to ensure all crawfish have access. The amount and frequency of feeding depend on the stocking density, pond size, and water temperature.

2. Natural Supplements:

- Green Plants: Fresh green plants like water lettuce, water hyacinth, and duckweed can be introduced into the pond as a natural supplement. Crawfish feed on these plants, which also contribute to the pond ecosystem by oxygenating water and absorbing excess nutrients.

- Animal Protein: Occasionally, small amounts of animal protein such as fish meal or shrimp pellets can be added to the diet to boost protein intake, especially during periods of rapid growth or reproductive phases.

Feeding Practices

- Monitoring Consumption: It's essential to monitor how much feed crawfish consume during feeding sessions. Overfeeding can lead to water quality issues due to increased organic matter decomposition, while underfeeding may stunt growth and reduce overall yield.

- Feeding Frequency: Typically, feeding is done once or twice a day, depending on the pond's productivity and the growth stage of the crawfish. In colder months or when crawfish are less active, feeding frequency may be reduced.

- Adjusting to Seasonal Changes: Feed requirements vary with seasonal changes in water temperature and daylight hours. During colder months, crawfish metabolism slows down, requiring less feed compared to warmer months when they are more active.

Feeding Management Tips

- Uniform Distribution: Ensure feed pellets are evenly distributed across the pond surface to prevent crowding and competition during feeding.

- Observation and Adjustment: Regularly observe crawfish behavior and pond conditions to adjust feeding practices accordingly. Factors such as weather changes, water quality, and population density can influence feeding patterns.

- Quality Control: Use high-quality feed pellets from reputable suppliers to ensure crawfish receive essential nutrients without compromising water quality.

Economic Considerations

- Cost Efficiency: Balancing the cost of supplemental feeding with increased yield and growth rates is crucial for maximizing profitability in crawfish farming.

- Nutritional Balance: Opt for feeds that provide a balanced nutritional profile suitable for crawfish at different life stages, from juvenile growth to mature breeding adults.

Environmental Impact

- Sustainable Practices: Implement feeding strategies that minimize environmental impact, such as using biodegradable feed pellets and avoiding excessive feeding practices that can lead to eutrophication.

Future Directions

- Research and Innovation: Continued research into optimizing feed formulations and feeding strategies can further enhance the efficiency and sustainability of crawfish farming operations.

- Technological Advances: Explore the potential of automated feeding systems and precision aquaculture technologies to improve feed efficiency and reduce labor costs.

3.3 Maintaining Pond Ecosystem

Maintaining a balanced and healthy pond ecosystem is essential for successful crawfish farming. Vegetation management is a critical aspect of this, as it directly impacts water quality, habitat availability, and overall productivity. Here's a detailed guide on how to effectively manage vegetation in crawfish ponds:

3.3.1 Vegetation Management

Vegetation in crawfish ponds can be beneficial or detrimental depending on its type, density, and management practices. Proper vegetation management aims to maximize benefits while minimizing potential drawbacks.

1. Types of Vegetation

Understanding the types of vegetation present in your pond is the first step in effective management:

- Beneficial Vegetation:

  - Aquatic Plants: Include submerged plants like pondweeds and emergent plants such as cattails. These plants provide cover and refuge for crawfish, offer natural food sources, and contribute to oxygenation of the water.

  - Algae: While often considered a nuisance, controlled growth of algae can serve as a natural food source for crawfish and other aquatic organisms.

- Detrimental Vegetation:

  - Overgrowth: Excessive growth of aquatic plants can lead to oxygen depletion during decay, block sunlight penetration, and hinder crawfish movement.

- Invasive Species: Non-native plants can outcompete native vegetation, disrupt the natural balance of the ecosystem, and potentially harm crawfish populations.

2. Vegetation Management Practices

Implementing effective vegetation management practices requires a combination of strategies tailored to the specific conditions of your pond:

- Regular Monitoring: Conduct frequent assessments of vegetation growth and composition. Monitor changes in water quality indicators such as oxygen levels and pH, which can be influenced by vegetation dynamics.

- Selective Harvesting: Controlled removal of excess vegetation helps prevent overgrowth and maintains favorable conditions for crawfish. Use rakes, aquatic harvesters, or manual labor to carefully harvest aquatic plants as needed.

- Biological Control: Introduce natural predators or competitors of invasive plants, such as herbivorous fish species or specific aquatic insects. This biological approach can help manage vegetation growth without resorting to chemical treatments.

- Mechanical Control: Utilize mechanical methods like dredging or cutting to physically remove unwanted vegetation. Mechanical control is particularly effective for large-scale management and can be supplemented with selective harvesting.

- Chemical Treatments: When necessary, use herbicides approved for aquatic use to control invasive species or manage vegetation in targeted areas of the pond. Always follow manufacturer instructions and regulatory guidelines to minimize environmental impact.

3. Environmental Considerations

Consider the broader environmental implications of vegetation management:

- Nutrient Cycling: Aquatic plants play a crucial role in nutrient cycling by absorbing and releasing nutrients. Proper management ensures efficient nutrient utilization, which contributes to water quality and overall pond health.

- Habitat Preservation: Maintain diverse vegetation types to provide essential habitats for crawfish at different life stages. Vegetation serves as shelter, spawning grounds, and foraging areas, supporting the overall biodiversity of the pond ecosystem.

- Erosion Control: Root systems of aquatic plants help stabilize pond banks and prevent soil erosion. This is especially important in preventing sedimentation that can negatively impact water quality and crawfish habitat.

4. Integrated Management Approach

Achieving sustainable vegetation management requires an integrated approach that considers ecological principles alongside practical management techniques:

- Plan Ahead: Develop a vegetation management plan based on pond characteristics, crawfish farming goals, and ecological considerations.

- Monitor Continuously: Regularly monitor vegetation growth, water quality parameters, and crawfish behavior to adjust management strategies as needed.

- Educate and Train: Ensure that personnel involved in vegetation management are properly trained in identification, monitoring techniques, and safe application of management practices.

- Adapt to Conditions: Adapt management strategies based on seasonal changes, weather patterns, and responses of vegetation and crawfish populations.

By following these detailed steps and considerations, crawfish farmers can effectively manage vegetation in their ponds to promote a healthy ecosystem conducive to optimal crawfish growth and productivity.

3.3.2 Predator Control

Crawfish ponds are vulnerable to a variety of predators that can significantly reduce your crawfish population if not properly managed. These predators include birds, mammals, reptiles, amphibians, and even certain fish. Effective predator control involves identifying these threats, understanding their behavior and impact, and implementing strategies to mitigate their presence.

Common Predators of Crawfish

1. Birds:

   - Herons and Egrets: These wading birds are skilled at hunting in shallow waters, making them significant predators of crawfish. Their sharp beaks and excellent eyesight enable them to catch crawfish with ease.

   - Kingfishers: Known for their diving capabilities, kingfishers can spot crawfish from above and swoop down to catch them.

   - Ducks: While primarily herbivores, some duck species may consume crawfish, especially if other food sources are scarce.

2. Mammals:

- Raccoons: These nocturnal animals are adept at foraging in and around water bodies. They use their dexterous front paws to catch crawfish and can significantly deplete pond populations if not controlled.

- Otters: As aquatic predators, otters are highly efficient at hunting crawfish. They are strong swimmers and can cover large areas in search of food.

3. *Reptiles and Amphibians:*

- Snakes: Some water snakes prey on crawfish, especially when they are molting and more vulnerable.

- Turtles: Both snapping turtles and softshell turtles are known to consume crawfish. They can be particularly problematic in ponds with high turtle populations.

4. *Fish:*

- Bass and Catfish: Certain fish species, particularly largemouth bass and catfish, may prey on young crawfish. While adult crawfish are generally safe from these predators, juveniles are at significant risk.

Impact of Predators on Crawfish Population

Predators can have a substantial impact on crawfish populations, leading to reduced yields and financial losses. Predation can disrupt the pond ecosystem by:

- Reducing the number of breeding adults, thus affecting future generations.

- Increasing the stress levels of crawfish, which can lead to stunted growth and increased susceptibility to disease.

- Competing with crawfish for food, thereby reducing the availability of natural food sources.

Strategies for Predator Control

Effective predator control requires a combination of proactive measures and responsive actions. The following strategies can help manage predator populations and minimize their impact on your crawfish pond.

1. Habitat Management:

 - Vegetation Control: Managing vegetation around the pond can make the environment less attractive to predators. For instance, maintaining low grass and removing tall reeds can reduce cover for raccoons and other mammals.

 - Water Depth Management: Adjusting the water depth can deter wading birds. Birds like herons prefer shallow waters, so maintaining deeper areas in the pond can make it more difficult for them to hunt.

2. Physical Barriers and Deterrents:

 - Fencing: Installing fences around the pond can help keep out terrestrial predators such as raccoons and otters. Electric fencing can be particularly effective.

 - Netting: Covering the pond with netting can prevent birds from accessing the water. This method is especially useful during the breeding season when crawfish are most vulnerable.

 - Decoys and Scare Devices: Using decoys (such as fake owls or alligators) and scare devices (like motion-activated sprinklers or noisemakers) can deter birds and other predators.

3. Trapping and Removal:

 - Live Traps: Setting live traps for raccoons, otters, and other mammals can help reduce their numbers. Once captured, these animals can be relocated to a safe distance from the pond.

 - Fish Traps: Installing traps specifically designed for predatory fish can help control their population. Regularly checking and removing trapped fish is necessary to maintain effectiveness.

4. Population Monitoring and Control:

- Regular Monitoring: Keeping a close eye on predator activity around the pond is crucial. Regular patrols, especially during dawn and dusk when many predators are active, can help identify and address problems early.

- Use of Predatory Controls: In some cases, introducing natural predators of the problematic species can help control their population. For instance, introducing certain fish species that prey on smaller predatory fish can help balance the ecosystem.

5. Environmental Modifications:

- Shelter and Refuge: Providing shelters, such as PVC pipes or rock piles, can offer crawfish a place to hide from predators. This is especially important for young or molting crawfish, which are more vulnerable.

- Alternative Food Sources: Ensuring there are ample natural food sources for predators outside the pond area can help divert their attention away from your crawfish. This can involve managing nearby ecosystems to support prey species that predators prefer over crawfish.

6. Chemical and Biological Controls:

- Repellents: There are chemical repellents available that can deter certain predators. These should be used with caution to avoid impacting the crawfish or the pond ecosystem negatively.

- Biological Controls: Introducing certain bacteria or fungi that affect predator species can be another method of control. However, this approach requires careful management to avoid unintended consequences.

Integrated Predator Management

The most effective predator control strategy often involves an integrated approach that combines multiple methods. Integrated Predator Management (IPM) is a holistic approach that aims to control predator populations using a combination of biological, physical, and chemical methods, along with habitat modification and regular monitoring. This approach minimizes reliance on any single method, reducing the risk of predators developing resistance or adapting to control measures.

Case Studies and Examples

1. Case Study: Heron Deterrence in Louisiana Crawfish Ponds:

 - Challenge: A crawfish farmer in Louisiana faced significant losses due to heron predation.

 - Solution: The farmer implemented a combination of netting over the pond, decoys, and regular use of motion-activated sprinklers. Additionally, the pond depth was adjusted to make it less suitable for herons.

 - Outcome: The combined measures resulted in a noticeable reduction in heron activity and a subsequent increase in crawfish survival rates.

2. Case Study: Raccoon Control in Texas:

 - Challenge: A crawfish pond in Texas experienced high predation rates from raccoons, especially during the breeding season.

 - Solution: The farmer installed electric fencing around the pond and set up live traps to capture and relocate raccoons. Vegetation around the pond was also managed to reduce cover for the raccoons.

 - Outcome: These measures significantly reduced raccoon predation, leading to improved crawfish yields.

Conclusion

Effective predator control is a critical aspect of maintaining a healthy and productive crawfish pond ecosystem. By understanding the various predators that pose a threat to crawfish and implementing a combination of habitat management, physical barriers, trapping, monitoring, and integrated management strategies, farmers can protect their crawfish populations and ensure sustainable production. Regular evaluation and adaptation of these strategies are essential to address changing conditions and emerging threats, ultimately contributing to the long-term success of the crawfish farming operation.

PART IV
Crawfish Breeding and Growth

4.1 Breeding Techniques

Breeding crawfish effectively requires a comprehensive understanding of their biology, environment, and the best practices in aquaculture. This section delves into the critical techniques for breeding crawfish, starting with the selection of breeding stock, which is the foundation of a successful breeding program.

4.1.1 Selecting Breeding Stock

Selecting the right breeding stock is one of the most crucial steps in crawfish farming. The quality of the breeding stock directly influences the overall health, growth rate, and productivity of the subsequent generations. This section covers the essential criteria and methods for choosing the best breeding stock for a thriving crawfish farm.

1. Understanding the Importance of Genetic Diversity

Genetic diversity in breeding stock is vital for maintaining a healthy and resilient population. It helps prevent the risks associated with inbreeding, such as reduced fertility, slower growth rates, and increased susceptibility to diseases. Therefore, farmers should aim to source breeding stock from multiple locations or ensure that their breeding pairs are not closely related.

2. Health and Physical Condition

When selecting breeding stock, prioritize crawfish that exhibit excellent health and robust physical condition. Key indicators of healthy crawfish include:

- Active Movement: Healthy crawfish are active and exhibit strong, coordinated movements.

- Intact Appendages: Ensure that the crawfish have all their limbs and claws, as these are essential for feeding, defense, and reproduction.

- Shell Condition: The exoskeleton should be hard and free of any cracks, lesions, or signs of molting stress. A healthy shell indicates good nutrition and overall health.

- Absence of Disease Symptoms: Avoid crawfish that show signs of disease, such as unusual coloration, lethargy, or visible parasites.

3. Size and Age Considerations

The size and age of the crawfish are important factors in breeding stock selection. Typically, mature adults are preferred for breeding purposes. Here are some guidelines:

- Optimal Size: Select crawfish that are large and robust, as they are likely to have better reproductive capabilities.

- Age: Choose crawfish that have reached sexual maturity. While the exact age can vary by species, most crawfish reach maturity within 6-12 months. It is crucial to avoid using too young or too old crawfish, as their reproductive success may be compromised.

4. Reproductive Traits

Evaluate the reproductive traits of potential breeding stock to ensure high productivity:

- Females: Look for females with broad abdomens and well-developed swimmerets, which are essential for carrying and aerating eggs. Additionally, females should have a history of successful spawning if possible.

- Males: Select males with large, well-formed chelae (claws) and a visible pair of gonopods (reproductive appendages). These traits are indicative of a male's ability to successfully mate and fertilize eggs.

5. Behavioral Traits

Behavioral traits can also influence the success of breeding. Ideal breeding stock should exhibit:

- Aggressiveness and Defensiveness: While overly aggressive individuals can be problematic, a certain level of assertiveness is necessary for defending territory and successful mating.

- Feeding Behavior: Active and consistent feeding behavior is a good indicator of overall health and vitality, which are important for successful reproduction.

6. Environmental Adaptability

Crawfish that are well adapted to their environment tend to have better reproductive success. When selecting breeding stock, consider individuals that have demonstrated resilience to local conditions, such as water temperature, pH levels, and salinity. Adaptability can enhance the overall robustness of the population.

7. Source Verification

Where you obtain your breeding stock matters. It's advisable to:

- Purchase from Reputable Suppliers: Ensure that the breeding stock comes from reputable sources known for maintaining healthy and genetically diverse populations.

- Inspect Source Conditions: Whenever possible, inspect the conditions in which the breeding stock was raised to ensure they were not exposed to poor water quality, overcrowding, or other stressors.

8. Acclimation and Quarantine

Before introducing new breeding stock into your existing population, it is essential to acclimate and quarantine them to prevent the spread of diseases and ensure they adapt to your farm's conditions:

- Gradual Acclimation: Slowly acclimate the new crawfish to the water conditions of your farm to reduce stress.

- Quarantine Period: Implement a quarantine period of at least 2-4 weeks to monitor the new stock for any signs of disease or health issues.

Conclusion

Selecting high-quality breeding stock is a fundamental step in establishing a successful crawfish farming operation. By prioritizing genetic diversity, health, physical and reproductive traits, and ensuring proper acclimation and quarantine, farmers can build a strong foundation for a productive and resilient crawfish population. The careful selection of breeding stock not only enhances the immediate success of breeding efforts but also contributes to the long-term sustainability and profitability of the farm.

4.1.2 Breeding Season Management

Breeding season management is a crucial aspect of successful crawfish farming. Proper management ensures high breeding success rates, robust offspring, and optimal resource utilization. This section will cover the various elements necessary for effective breeding season management, including timing, environmental control, feeding, and handling of breeding stock.

Timing the Breeding Season

The timing of the breeding season is vital for maximizing reproductive success in crawfish farming. In general, the breeding season for most crawfish species occurs in the warmer months when water temperatures are favorable for breeding and larval development. For instance, in temperate regions, breeding typically takes place from late spring to early fall.

To optimize breeding, farmers should monitor water temperatures closely. Crawfish generally begin breeding when water temperatures consistently remain between 18°C (64°F) and 28°C (82°F). However, this can vary slightly depending on the species and local conditions. Ensuring that breeding starts at the right time can significantly impact the success rate and health of the offspring.

Environmental Control

Environmental factors play a pivotal role in the breeding process of crawfish. Proper management of water quality, habitat structure, and overall pond conditions is essential for promoting successful breeding.

1. Water Quality: Maintaining optimal water quality is critical. Parameters such as pH, dissolved oxygen, and ammonia levels must be regularly monitored and kept within acceptable ranges. The pH level should ideally be between 6.5 and 8.5. Dissolved oxygen levels should be above 5 ppm, and ammonia levels should be kept as low as possible, preferably below 0.5 ppm.

2. Habitat Structure: Crawfish require a suitable habitat for breeding, which includes ample hiding places and substrate for burrowing. Providing artificial shelters, such as PVC pipes or clay pots, can help mimic natural conditions and encourage breeding. The substrate should be soft and muddy, allowing crawfish to dig and create burrows.

3. Water Flow and Depth: Moderate water flow and appropriate water depth are also important. A depth of 30-60 cm (12-24 inches) is generally recommended for breeding ponds. Ensuring a gentle flow of water helps maintain water quality and provides a dynamic environment that supports breeding activities.

Feeding During Breeding Season

Proper nutrition is crucial during the breeding season to ensure the health and reproductive success of the breeding stock. Crawfish have increased nutritional needs during this time, and providing a balanced diet is essential.

1. Nutrient-Rich Diet: Crawfish should be fed a diet rich in protein and essential nutrients. Commercially available crawfish feed is formulated to meet these needs, but supplementary feeding with natural food sources, such as aquatic plants, insects, and detritus, can enhance their diet.

2. Feeding Frequency: During the breeding season, feeding frequency should be increased to ensure that the crawfish have constant access to food. Feeding twice a day, in the morning and evening, is generally recommended.

3. Supplemental Feeding: In addition to commercial feed, supplemental feeding with high-protein foods, such as fish meal or soybean meal, can be beneficial. These supplements can help improve the reproductive capacity and overall health of the breeding stock.

Handling Breeding Stock

Proper handling of breeding stock is essential to minimize stress and injury, which can negatively impact breeding success. The following practices are recommended for handling crawfish during the breeding season:

1. Minimize Handling: Handling should be minimized to reduce stress. When handling is necessary, it should be done gently and efficiently. Crawfish should be lifted carefully to avoid damage to their delicate exoskeletons.

2. Segregation of Breeding Stock: Segregating breeding stock from the general population can help reduce competition and aggression, providing a more conducive environment for breeding. Breeding pairs or groups can be placed in separate enclosures or breeding tanks.

3. Monitoring and Care: Regular monitoring of breeding stock is essential to ensure their health and well-being. Observations should include checking for signs of stress, injury, or disease. Any issues should be addressed promptly to maintain optimal breeding conditions.

Disease Prevention and Management

Disease prevention is a critical aspect of breeding season management. Crawfish are susceptible to various diseases, and outbreaks can significantly impact breeding success. Implementing effective disease prevention and management strategies is essential for maintaining a healthy breeding population.

1. Regular Health Checks: Conduct regular health checks to identify and address any potential health issues early. This includes inspecting crawfish for signs of disease, such as discoloration, abnormal behavior, or lesions.

2. Biosecurity Measures: Implement biosecurity measures to prevent the introduction and spread of diseases. This includes controlling access to breeding areas, disinfecting equipment, and ensuring that any new stock is quarantined before introduction.

3. Disease Management Protocols: Develop and implement disease management protocols to address any outbreaks that may occur. This should include isolation of affected individuals, treatment with appropriate medications, and thorough cleaning and disinfection of breeding areas.

Record Keeping and Data Analysis

Maintaining detailed records and analyzing data is crucial for successful breeding season management. Accurate records help farmers track breeding success, identify trends, and make informed decisions for future breeding seasons.

1. Breeding Logs: Keep detailed logs of breeding activities, including dates of pairing, number of eggs laid, and hatching success rates. This information can help identify the most successful breeding pairs and optimal breeding conditions.

2. Health Records: Maintain health records for breeding stock, including any treatments administered and outcomes. This can help identify patterns and potential health issues that may affect breeding success.

3. Environmental Data: Record environmental data, such as water temperature, pH, and dissolved oxygen levels, throughout the breeding season. Analyzing this data can help identify optimal environmental conditions for breeding.

4. Feeding Records: Track feeding schedules and types of feed provided. This information can help assess the effectiveness of different feeding strategies and their impact on breeding success.

By effectively managing the breeding season through careful timing, environmental control, proper nutrition, and diligent handling of breeding stock, crawfish farmers can significantly enhance their breeding success and overall productivity. Implementing disease prevention and management strategies, along with maintaining detailed records, further ensures the health and viability of the breeding population, leading to a sustainable and profitable crawfish farming operation.

4.2 Growth Stages

4.2.1 Juvenile Stage

The juvenile stage is a critical period in the lifecycle of crawfish, marked by rapid growth and significant development. Understanding the intricacies of this stage is essential for successful crawfish farming, as it sets the foundation for the health and productivity of the adult population. This section delves into the physiological changes, environmental needs, and management practices essential for optimizing the growth of juvenile crawfish.

Physiological Development

During the juvenile stage, crawfish undergo significant physiological changes. After hatching from eggs, they progress from larvae to juveniles. This transition is characterized by several molts, where the crawfish shed their exoskeleton to accommodate their growing bodies. Molting is a vulnerable time for juveniles as their new exoskeletons are soft, making them susceptible to predation and environmental stressors. Providing a safe habitat with plenty of hiding places, such as vegetation or artificial structures, is crucial during this period.

Nutritional Requirements

Juvenile crawfish have high nutritional needs to support their rapid growth. A balanced diet rich in proteins, fats, and essential minerals is vital. In natural settings, juveniles feed on a variety of plant matter, small invertebrates, and detritus. In a farming environment, this diet can be supplemented with high-quality commercial feeds specifically formulated for juvenile crustaceans. These feeds should contain a mix of animal and plant proteins, along with essential fatty acids and vitamins. Regular feeding schedules and monitoring of food intake help ensure that juveniles receive adequate nutrition, promoting healthy growth and development.

Environmental Conditions

Maintaining optimal environmental conditions is critical for the survival and growth of juvenile crawfish. Key factors include water quality, temperature, and oxygen levels. Juveniles thrive in clean, well-oxygenated water with stable temperatures. Sudden changes in water quality or temperature can stress juveniles, leading to stunted growth or increased mortality rates.

- Water Quality: Regular monitoring and management of water parameters such as pH, ammonia, nitrite, and nitrate levels are essential. Installing a good filtration system and performing regular water changes can help maintain optimal water quality.

- Temperature: Juvenile crawfish grow best in water temperatures ranging from 70°F to 80°F (21°C to 27°C). Consistent temperature regulation can be achieved using heaters or chillers, depending on the climate and season.

- Oxygen Levels: Adequate dissolved oxygen levels are crucial. Aeration devices like air stones or diffusers can help maintain sufficient oxygen in the water, especially in densely populated tanks or ponds.

Habitat and Space Requirements

Providing an appropriate habitat is crucial for the well-being and growth of juvenile crawfish. Overcrowding can lead to increased competition for food and space, resulting in slower growth rates and higher stress levels. It is important to maintain an appropriate stocking density to ensure that each juvenile has enough space to grow and access resources.

- Stocking Density: The recommended stocking density for juvenile crawfish varies depending on the farming system and species. Generally, a lower density is preferred to minimize competition and stress.

- Habitat Enrichment: Introducing habitat structures like PVC pipes, clay pots, or specially designed shelters can provide juveniles with hiding places and reduce aggressive interactions. Vegetation, both natural and artificial, can also enhance the habitat by providing cover and additional foraging opportunities.

Health Monitoring and Disease Management

Regular health monitoring is essential to detect early signs of disease or stress in juvenile crawfish. Common health issues in juveniles include fungal infections, bacterial diseases, and parasites. Implementing a robust health management plan can help prevent and control these issues.

- Regular Health Checks: Conducting routine inspections of juveniles for signs of disease, such as discoloration, unusual behavior, or physical deformities, is crucial. Healthy juveniles should be active and exhibit normal feeding behavior.

- Disease Prevention: Preventive measures include maintaining good water quality, providing a balanced diet, and avoiding overcrowding. Quarantining new stock before introducing them to the main population can also prevent the spread of diseases.

- Treatment: If a disease outbreak occurs, prompt treatment is necessary. Depending on the disease, treatments may include medicated feeds, water treatments, or antibiotics. Consulting with a veterinarian or aquaculture specialist can help determine the best course of action.

Growth Monitoring and Record Keeping

Tracking the growth of juvenile crawfish is important for assessing the effectiveness of farming practices and making necessary adjustments. Regular measurements of weight and length can provide valuable data on growth rates and overall health.

- Growth Measurements: Periodically measuring a sample of juveniles can help monitor growth progress. Comparing these measurements against expected growth rates can indicate whether juveniles are developing as anticipated.

- Record Keeping: Maintaining detailed records of growth data, feeding schedules, water quality parameters, and health observations is essential for effective farm management. These records can help identify trends, diagnose issues, and improve future farming practices.

Transition to Adult Stage

The juvenile stage concludes when crawfish reach a size where they can be considered young adults. This transition is marked by a decrease in molting frequency and a shift in dietary and habitat needs. Preparing juveniles for this transition involves gradually adjusting their environment and diet to meet the needs of adult crawfish.

- Dietary Adjustments: As juveniles grow, their nutritional needs change. Gradually introducing adult-formulated feeds while continuing to provide a varied diet can help ease this transition.

- Habitat Modifications: Adjusting the habitat to accommodate larger crawfish is important. This may involve increasing the size of shelters and ensuring that the habitat remains enriched and conducive to healthy growth.

In conclusion, the juvenile stage is a pivotal period in the lifecycle of crawfish, requiring careful management and attention to detail. By providing optimal environmental conditions, a balanced diet, adequate space, and vigilant health monitoring, farmers can ensure the healthy growth and development of juvenile crawfish, laying the foundation for a productive and sustainable farming operation.

4.2.2 Adult Stage

The adult stage in crawfish farming is a critical period where the organisms reach maturity and are prepared for either breeding or harvest. Understanding this stage involves recognizing the biological, environmental, and management factors that influence the development, health, and productivity of adult crawfish. This section will delve into various aspects of the adult stage, including physical development, optimal conditions for growth, feeding practices, health monitoring, and overall management strategies to maximize yield and ensure sustainable farming practices.

Physical Development and Characteristics of Adult Crawfish

Crawfish typically reach adulthood approximately 3-6 months after hatching, depending on species, environmental conditions, and nutrition. Adult crawfish are characterized by their fully developed exoskeleton, which provides protection and support. They exhibit pronounced sexual dimorphism, meaning males and females can be distinguished by specific physical traits. Males generally have larger and more robust chelae (claws), while females possess broader abdomens to accommodate egg carrying.

Adult crawfish undergo periodic molting, shedding their exoskeleton to allow for further growth. This process is crucial for their development but also leaves them vulnerable to predation and environmental stress. Proper management during molting periods is essential to ensure the safety and continued growth of the crawfish.

Optimal Environmental Conditions for Adult Crawfish

Maintaining the right environmental conditions is crucial for the health and productivity of adult crawfish. Key factors include water quality, temperature, habitat structure, and population density.

1. Water Quality: Clean, well-oxygenated water with appropriate pH levels (typically between 6.5 and 8.5) is essential. Regular monitoring of ammonia, nitrite, and nitrate levels is necessary to prevent toxic build-up. Water hardness and alkalinity should also be maintained within optimal ranges to support exoskeleton development and molting.

2. Temperature: Crawfish are ectothermic, meaning their body temperature is regulated by the surrounding environment. Ideal temperatures for adult crawfish range from 20°C to 28°C (68°F to 82°F). Temperatures outside this range can stress the crawfish, affecting their growth and reproduction.

3. Habitat Structure: Providing a habitat with ample hiding spaces, such as rocks, vegetation, and artificial shelters, reduces stress and cannibalism among adult crawfish. These structures also offer protection during molting periods.

4. Population Density: Overcrowding can lead to increased competition for resources, higher stress levels, and greater susceptibility to disease. Optimal stocking densities vary but generally range from 2 to 4 crawfish per square meter. Regular monitoring and adjustments are necessary to maintain healthy population levels.

Feeding Practices for Adult Crawfish

Nutrition plays a vital role in the growth and health of adult crawfish. A balanced diet ensures proper development, supports molting, and enhances reproductive success. Key considerations for feeding adult crawfish include:

1. Diet Composition: Adult crawfish are omnivorous, consuming both plant and animal matter. A balanced diet typically includes commercial pellets formulated specifically for crustaceans, supplemented with fresh or dried plant material (such as vegetables and aquatic plants) and animal protein (such as fish or shrimp).

2. Feeding Frequency and Quantity: Feeding practices should align with the natural foraging behavior of crawfish. Offering food once or twice daily is generally sufficient. The quantity of food provided should be enough to meet the nutritional needs without causing excess waste, which can degrade water quality.

3. Feeding Techniques: Distributing food evenly throughout the habitat encourages natural foraging behavior and reduces competition among individuals. Monitoring feeding habits can help adjust the amount and type of food offered, ensuring optimal nutrition and minimizing waste.

Health Monitoring and Disease Management

Regular health monitoring is essential to identify and address potential issues promptly. Common health checks include observing behavior, physical condition, and any signs of disease or stress.

1. Behavioral Observation: Healthy adult crawfish are typically active, foraging for food, and displaying normal social interactions. Changes in behavior, such as lethargy, reduced feeding, or increased aggression, can indicate health problems or environmental stress.

2. Physical Condition: Regularly inspecting crawfish for physical abnormalities, such as discoloration, lesions, or deformities, helps detect early signs of disease or injury. Molting problems, such as incomplete molts or difficulty shedding the exoskeleton, should also be monitored closely.

3. Disease Prevention and Management: Common diseases affecting adult crawfish include bacterial infections, fungal infections, and parasitic infestations. Preventative measures include maintaining optimal water quality, providing a balanced diet, and minimizing stress through proper habitat management. If disease outbreaks occur, prompt isolation of affected individuals and appropriate treatment (such as antibiotics or antifungal medications) are necessary to prevent spread.

Reproductive Management of Adult Crawfish

For farmers focusing on breeding, managing the reproductive cycle of adult crawfish is crucial. Key aspects include:

1. Mating Behavior: Understanding the mating behavior of crawfish, such as courtship rituals and copulation processes, helps create conducive conditions for successful reproduction. Providing suitable habitats with hiding spaces and controlled population densities encourages natural mating.

2. Egg Development and Incubation: After mating, females carry fertilized eggs on their swimmerets (appendages under the abdomen). Ensuring optimal environmental conditions, such as stable water quality and temperature, supports egg development and hatching success.

3. Juvenile Care: Once eggs hatch, juvenile crawfish require specific care to ensure their survival and growth. Proper nutrition, habitat structure, and protection from predators (including adult crawfish) are essential during the early stages of life.

Harvesting and Post-Harvest Management

When adult crawfish reach market size, harvesting becomes the primary focus. Effective harvesting techniques and post-harvest management ensure high-quality products and sustainable farming practices.

1. Harvesting Techniques: Various methods, such as baited traps or seine nets, are used to capture adult crawfish. Harvesting should be done carefully to minimize stress and injury. Regular harvesting helps maintain optimal population densities and promotes continued growth and reproduction.

2. Post-Harvest Handling: Proper handling and storage of harvested crawfish are crucial to maintain quality and freshness. Immediately after harvesting, crawfish should be sorted,

cleaned, and stored in cool, well-ventilated containers. Minimizing handling time and exposure to high temperatures reduces stress and mortality.

3. Market Preparation: Preparing crawfish for market involves grading, packaging, and transportation. Grading ensures that crawfish meet size and quality standards. Packaging should protect crawfish from damage and dehydration during transport. Maintaining a cool, moist environment during transportation preserves the quality and extends the shelf life of the product.

Sustainable Farming Practices

Sustainability in crawfish farming involves practices that ensure long-term productivity and minimal environmental impact. Key strategies include:

1. Resource Management: Efficient use of water, feed, and other resources reduces waste and environmental impact. Implementing recirculating water systems and sustainable feed sources supports eco-friendly farming practices.

2. Biodiversity Conservation: Protecting and promoting biodiversity within and around the farming area enhances ecosystem health and resilience. Preserving natural habitats and minimizing pollution contribute to sustainable farming.

3. Integrated Pest Management: Implementing integrated pest management (IPM) strategies reduces reliance on chemical treatments and promotes natural pest control methods. IPM includes biological control, habitat management, and selective use of pesticides.

4. Community Engagement: Engaging with local communities and stakeholders promotes sustainable practices and supports social and economic well-being. Sharing knowledge,

resources, and benefits with the community fosters a positive relationship and mutual support.

Conclusion

The adult stage in crawfish farming is a complex and dynamic period requiring careful management to ensure the health, productivity, and sustainability of the farm. By understanding the physical development, optimal conditions, feeding practices, health monitoring, reproductive management, and sustainable practices, farmers can successfully navigate this stage and achieve high yields of quality crawfish. Through continuous learning and adaptation, crawfish farmers can contribute to the growth and sustainability of the aquaculture industry, providing a valuable source of food and income for communities worldwide.

4.3 Monitoring Growth and Health

4.3.1 Regular Health Checks

Regular health checks are vital in crawfish farming to ensure the wellbeing and optimal growth of the crawfish population. These checks help in early detection of diseases, nutritional deficiencies, and other health-related issues that can impact the productivity and sustainability of the farming operation. Implementing a systematic and thorough health monitoring routine will contribute significantly to maintaining a healthy and thriving crawfish farm.

Importance of Regular Health Checks

1. Early Detection of Diseases:

   Early identification of diseases can prevent widespread outbreaks. Regular health checks allow farmers to spot early symptoms of illnesses, which can be critical in managing and controlling diseases before they cause significant damage to the population.

2. Monitoring Growth Rates:

   Regular assessments of growth rates help farmers ensure that the crawfish are developing at the expected pace. Deviations from the standard growth patterns can indicate health issues or environmental problems that need addressing.

3. Ensuring Optimal Nutrition:

   Health checks can help identify nutritional deficiencies. Observing the physical condition of the crawfish, such as shell hardness and coloration, provides insights into whether the crawfish are receiving the necessary nutrients.

4. Environmental Monitoring:

Regular health checks often include monitoring water quality parameters such as pH, temperature, dissolved oxygen, and ammonia levels. These environmental factors significantly affect the health and growth of crawfish, and maintaining them within optimal ranges is crucial.

Procedures for Regular Health Checks

1. Visual Inspections:

Conduct visual inspections of the crawfish for any signs of physical abnormalities, such as lesions, discolorations, or unusual behaviors. Healthy crawfish should have a robust shell, clear eyes, and exhibit active movement.

2. Sampling and Testing:

Periodically sample a portion of the crawfish population for closer examination. This can include checking for parasites, bacterial infections, or any other signs of illness. Laboratory tests may be necessary to diagnose specific conditions accurately.

3. Weight and Size Measurements:

Regularly measure the weight and size of the crawfish to monitor their growth. Keeping detailed records of these measurements helps track growth trends and identify any deviations that may indicate health issues.

4. Behavioral Observations:

Observe the behavior of the crawfish. Healthy crawfish are typically active and responsive. Lethargy, erratic swimming, or unusual hiding behaviors can be signs of health problems.

5. Water Quality Testing:

Test the water regularly for key parameters such as pH, temperature, dissolved oxygen, ammonia, nitrites, and nitrates. Poor water quality can lead to stress and disease in crawfish, so maintaining optimal water conditions is essential for their health.

Implementing a Health Monitoring Schedule

1. Daily Checks:

   - Perform quick visual inspections of the crawfish and check their behavior.

   - Monitor water temperature and dissolved oxygen levels.

   - Remove any dead or visibly ill crawfish to prevent the spread of disease.

2. Weekly Checks:

   - Conduct more detailed visual inspections and record any abnormalities.

   - Test water quality parameters, including pH, ammonia, nitrites, and nitrates.

   - Measure and record the size and weight of a sample group of crawfish.

3. Monthly Checks:

   - Perform comprehensive health checks, including sampling for laboratory tests if necessary.

   - Review and analyze growth data to ensure the crawfish are developing as expected.

   - Inspect and clean the habitat to prevent the buildup of waste and harmful bacteria.

4. Seasonal Checks:

   - Assess the overall health of the entire population, especially before and after significant environmental changes such as seasonal temperature shifts.

- Adjust feeding practices and nutritional supplements based on the health and growth data collected.

Common Health Issues in Crawfish

1. Shell Disease:

This bacterial infection causes lesions on the shell, leading to weakened defenses and increased mortality. Regular health checks can identify early signs of shell disease, allowing for prompt treatment.

2. Parasitic Infections:

Parasites can cause significant health problems and reduce the growth rate of crawfish. Regular inspections and laboratory tests help detect and manage parasitic infections.

3. Nutritional Deficiencies:

Insufficient nutrition can lead to weak shells, poor growth, and increased susceptibility to diseases. Monitoring the physical condition of the crawfish helps ensure they receive adequate nutrition.

4. Environmental Stress:

Poor water quality, overcrowding, and inadequate habitat conditions can stress crawfish, making them more vulnerable to diseases. Regular environmental monitoring and adjustments help mitigate stress factors.

Best Practices for Health Management

1. Maintain Optimal Water Quality:

Regularly test and adjust water quality parameters to ensure they remain within the optimal range for crawfish health. Use appropriate filtration and aeration systems to maintain clean and oxygen-rich water.

2. Provide Balanced Nutrition:

Ensure that the crawfish receive a balanced diet with all the necessary nutrients. Use high-quality commercial feeds and supplement with natural food sources when possible.

3. Implement Biosecurity Measures:

Prevent the introduction of diseases by implementing strict biosecurity measures. This includes controlling access to the farm, disinfecting equipment, and quarantining new stock before introducing them to the main population.

4. Regular Cleaning and Maintenance:

Keep the crawfish habitat clean and free of waste. Regularly remove uneaten food and debris to prevent the buildup of harmful bacteria and parasites.

5. Educate and Train Staff:

Ensure that all staff members are trained in proper health monitoring techniques and understand the importance of regular health checks. Educated and vigilant staff are crucial for early disease detection and effective health management.

Conclusion

Regular health checks are an indispensable part of crawfish farming, playing a crucial role in maintaining the health, growth, and productivity of the crawfish population. By implementing a systematic health monitoring routine, farmers can detect and address health issues early, ensure optimal nutrition, and maintain a healthy environment for the crawfish. This proactive approach not only improves the wellbeing of the crawfish but also enhances the overall success and sustainability of the farming operation.

4.3.2 Disease Prevention and Management

Crawfish farming, like any other aquaculture endeavor, requires meticulous attention to the health of the stock to ensure a successful and profitable harvest. One of the most critical aspects of this is disease prevention and management. Understanding the common diseases that affect crawfish, their symptoms, and the preventive measures that can be taken is crucial for any beginner crawfish farmer.

Understanding Common Diseases

Crawfish are susceptible to various diseases, most of which are caused by bacteria, fungi, parasites, and viruses. Some of the most common diseases include:

1. White Spot Syndrome Virus (WSSV): This viral disease is highly contagious and can cause significant mortality in crawfish populations. Infected crawfish show white spots on their exoskeleton and may exhibit lethargy and a reduced appetite.

2. Crayfish Plague (Aphanomyces astaci): Caused by a water mold, this disease is devastating to crawfish. Symptoms include a loss of coordination, color changes, and eventual death. It is highly infectious and often fatal.

3. Bacterial Infections: Various bacteria can infect crawfish, leading to conditions such as shell disease, which causes black spots on the exoskeleton, and vibriosis, which can result in septicemia and high mortality rates.

4. Parasitic Infections: Crawfish can host various parasites, including protozoa and helminths. These parasites can cause significant damage to the internal organs and affect the overall health of the crawfish.

5. Fungal Infections: Fungi can infect the crawfish, particularly when they are stressed or injured. These infections can lead to issues like cotton wool disease, which appears as white, cotton-like growths on the body.

Symptoms and Diagnosis

Recognizing the symptoms of these diseases early is essential for effective management. Some general signs of disease in crawfish include:

- Lethargy and reduced movement

- Loss of appetite

- Visible spots or lesions on the exoskeleton

- Abnormal color changes

- Uncoordinated swimming or movements

- Presence of parasites or fungal growths on the body

Diagnosing the specific disease often requires laboratory tests, including microscopic examinations and microbial cultures. In some cases, molecular methods like PCR (Polymerase Chain Reaction) are used to detect specific pathogens, such as viruses.

Preventive Measures

Prevention is always better than cure, and there are several strategies that crawfish farmers can implement to minimize the risk of disease outbreaks:

1. Water Quality Management: Maintaining optimal water quality is crucial for the health of the crawfish. Regular monitoring of parameters like pH, temperature, dissolved oxygen, and ammonia levels helps ensure a suitable environment for the crawfish.

2. Stocking Density: Overstocking can lead to increased stress and a higher risk of disease outbreaks. Farmers should adhere to recommended stocking densities to prevent overcrowding and ensure adequate space for each crawfish.

3. Biosecurity Measures: Implementing strict biosecurity protocols can prevent the introduction and spread of diseases. This includes controlling access to the farming site, disinfecting equipment, and ensuring that any new stock is disease-free before introduction.

4. Nutrition and Feeding: Providing a balanced diet that meets the nutritional needs of the crawfish can enhance their immune system and overall health. Farmers should use high-quality feeds and avoid overfeeding, which can deteriorate water quality.

5. Regular Health Checks: Routine health inspections help in early detection of potential health issues. Farmers should regularly observe their stock for any signs of disease and take prompt action if any abnormalities are noticed.

6. Quarantine Procedures: New or sick crawfish should be quarantined before being introduced to the main population. This practice helps in isolating potential carriers of disease and reduces the risk of widespread infection.

Disease Management

Despite the best preventive measures, disease outbreaks can still occur. Effective management of these outbreaks is essential to minimize losses and restore the health of the crawfish population. The following steps can be taken for disease management:

1. Isolation of Affected Stock: Infected crawfish should be immediately isolated to prevent the spread of the disease. This involves setting up a separate quarantine area where the affected individuals can be treated.

2. Medication and Treatment: Depending on the type of disease, appropriate treatments should be administered. For bacterial infections, antibiotics may be used, whereas antifungal treatments are necessary for fungal infections. It is important to use medications as per veterinary guidance to avoid resistance and ensure effectiveness.

3. Improving Environmental Conditions: Enhancing water quality and reducing stressors can aid in the recovery of diseased crawfish. This might involve increasing aeration, adjusting water parameters, and providing a cleaner habitat.

4. Probiotics and Immunostimulants: Adding probiotics to the water or feed can help in maintaining a healthy gut flora in crawfish, which boosts their immune system. Immunostimulants can also be used to enhance the natural defenses of the crawfish against infections.

5. Regular Monitoring and Documentation: Keeping detailed records of disease incidents, treatments administered, and outcomes is essential. This data helps in understanding the patterns of disease occurrence and evaluating the effectiveness of the management strategies implemented.

6. Collaboration with Veterinarians and Experts: Seeking the advice of aquaculture veterinarians and disease experts can provide valuable insights and recommendations for

dealing with specific disease issues. Collaborating with professionals ensures that the best practices are followed in disease management.

Case Studies and Examples

To illustrate the importance of disease prevention and management, let's consider a few case studies from successful crawfish farms:

1. Case Study 1: White Spot Syndrome Virus Outbreak:

   - A crawfish farm experienced a sudden outbreak of WSSV, leading to high mortality rates.

   - The farmer immediately isolated the affected ponds and improved biosecurity measures.

   - Water quality was optimized, and infected stock was treated with antiviral agents.

   - After rigorous management, the outbreak was controlled, and the farm implemented stricter biosecurity protocols to prevent future occurrences.

2. Case Study 2: Bacterial Shell Disease:

   - A farmer noticed black spots on the shells of several crawfish, indicating a bacterial infection.

   - The infected individuals were isolated, and antibiotics were administered as per veterinary advice.

   - The farm's water quality management practices were reviewed and improved, reducing the recurrence of the disease.

   - Regular health checks and the use of probiotics helped in maintaining the overall health of the crawfish population.

3. Case Study 3: Parasitic Infestation:

- A crawfish farm detected a parasitic infestation that was affecting the growth and health of the stock.

- The affected crawfish were quarantined, and antiparasitic treatments were administered.

- The farm improved its screening processes for new stock to prevent future infestations.

- Enhanced biosecurity and regular monitoring ensured that the farm remained free from significant parasitic issues.

Conclusion

Disease prevention and management are integral to the success of crawfish farming. By understanding the common diseases that affect crawfish, recognizing their symptoms, and implementing effective preventive measures, farmers can significantly reduce the risk of disease outbreaks. In the event of an outbreak, prompt and effective management is crucial to minimize losses and restore the health of the crawfish population. Through diligent monitoring, maintaining optimal environmental conditions, and adhering to biosecurity protocols, crawfish farmers can ensure the health and productivity of their stock, leading to a successful and sustainable farming operation.

PART V
Harvesting Crawfish

5.1 Harvesting Techniques

Harvesting techniques in crawfish farming are crucial to maximizing yield and ensuring the quality of the catch. Efficient harvesting not only determines the profitability of the operation but also impacts the sustainability of the crawfish population and the overall ecosystem. This section will delve into various methods used in harvesting crawfish, with a primary focus on trapping methods.

5.1.1 Trapping Methods

Trapping methods are among the most commonly used techniques for harvesting crawfish. These methods are effective, relatively easy to implement, and can be scaled up or down depending on the size of the farming operation. Trapping involves the use of devices specifically designed to capture crawfish without causing them significant harm. This section explores different types of traps, their design, deployment strategies, and best practices for maximizing efficiency.

Types of Traps

1. Pyramid Traps:

   - Design: Pyramid traps are shaped like a three-sided pyramid with an open bottom. They are typically made from wire mesh or plastic netting.

- Functionality: The open bottom allows crawfish to enter the trap from the pond or water body. Once inside, the pyramid shape and the design of the entrance make it difficult for them to escape.

- Advantages: Pyramid traps are easy to set up and can capture a significant number of crawfish. They are also relatively inexpensive and durable.

2. Box Traps:

- Design: Box traps are rectangular and have multiple entrance funnels. They are constructed from wire mesh and often have a bait compartment in the center.

- Functionality: Crawfish are lured into the trap by the bait. The funnels allow crawfish to enter but make it challenging for them to exit.

- Advantages: Box traps can hold a larger quantity of crawfish compared to pyramid traps. They are also effective in different water conditions.

3. Cone Traps:

- Design: Cone traps are conical in shape with an entrance at the wide base and a narrow top where crawfish are collected.

- Functionality: Crawfish enter through the wide base and move upwards towards the bait, becoming trapped at the narrow top.

- Advantages: Cone traps are efficient in capturing crawfish and can be easily transported and stored.

Deployment Strategies

1. Site Selection:

- Optimal Locations: Traps should be placed in areas with high crawfish activity, such as near vegetation, under logs, and in shallow waters where crawfish tend to feed and hide.

- Environmental Factors: Consider factors such as water temperature, depth, and flow when selecting trap locations. Crawfish are more active in warmer temperatures and tend to avoid areas with strong currents.

2. Baiting:

   - Types of Bait: Common baits include fish parts, commercial crawfish bait, and vegetable matter. The choice of bait can significantly impact the effectiveness of the traps.

   - Bait Placement: Place bait securely within the trap to ensure it remains effective throughout the trapping period. Some traps have dedicated bait compartments to keep the bait in place.

3. Setting the Traps:

   - Frequency: Traps should be checked and emptied regularly, typically every 24-48 hours. Frequent checks prevent overfilling and ensure the crawfish remain alive and in good condition.

   - Techniques: Gently lower the traps into the water to avoid scaring away the crawfish. Ensure the traps are fully submerged but not buried in the mud or sediment.

4. Monitoring and Maintenance:

   - Regular Inspections: Check traps for damage and repair any holes or weak spots to maintain their effectiveness.

   - Environmental Impact: Monitor the impact of trapping on the crawfish population and the surrounding ecosystem. Avoid overharvesting to ensure sustainability.

Best Practices for Maximizing Efficiency

1. Timing:

- Optimal Harvesting Seasons: Crawfish harvesting is typically more productive during certain seasons. Late spring to early summer is often the peak period for crawfish activity.

- Daily Timing: Crawfish are nocturnal and more active at night. Setting and checking traps during the evening and early morning can yield better results.

2. Rotation and Variation:

- Rotating Trap Locations: Regularly move traps to different locations within the farming area to cover a broader range and prevent local depletion of crawfish.

- Varying Trap Types: Use a combination of different trap types to target crawfish of varying sizes and behaviors.

3. Record Keeping:

- Data Collection: Keep detailed records of trap locations, bait types, catch numbers, and environmental conditions. This data can help identify trends and improve harvesting strategies.

- Analysis: Analyze the data to determine the most effective practices and make informed decisions about future trapping efforts.

4. Sustainability:

- Conservation Measures: Implement measures to protect young and breeding crawfish. Consider using traps with escape gaps for smaller, immature crawfish to promote population growth.

- Habitat Management: Maintain a healthy habitat by controlling water quality, vegetation, and other environmental factors that support crawfish populations.

Challenges and Solutions

1. Environmental Variability:

- Challenge: Changes in water temperature, flow, and quality can affect crawfish behavior and trap effectiveness.

- Solution: Regularly monitor environmental conditions and adjust trapping strategies accordingly. Implement adaptive management practices to respond to changing conditions.

2. Predation and Competition:

- Challenge: Predators and competing species can reduce the number of crawfish captured in traps.

- Solution: Use traps designed to minimize entry by non-target species. Place traps in locations less accessible to predators.

3. Trap Theft and Vandalism:

- Challenge: Theft and vandalism of traps can be a significant issue, especially in areas with public access.

- Solution: Mark traps with identification tags and keep records of trap locations. Consider using surveillance or security measures in high-risk areas.

4. Labor and Resource Management:

- Challenge: Efficiently managing the labor and resources required for trapping can be demanding, especially in large-scale operations.

- Solution: Optimize trapping schedules and workflows to balance labor costs with productivity. Invest in durable, high-quality traps to reduce maintenance and replacement costs.

Technological Advancements

1. Automated Traps:

- Innovation: Development of automated traps that can capture and hold crawfish without frequent manual intervention.

- Advantages: Reduces labor requirements and increases efficiency, especially in large-scale operations.

2. Remote Monitoring:

- Innovation: Use of remote monitoring systems to track trap locations, catches, and environmental conditions in real-time.

- Advantages: Provides valuable data for optimizing trapping strategies and reduces the need for frequent on-site inspections.

3. Biodegradable Traps:

- Innovation: Development of biodegradable traps that reduce environmental impact and prevent long-term pollution if lost or abandoned.

- Advantages: Supports sustainable farming practices and minimizes harm to the ecosystem.

Conclusion

Trapping methods are a cornerstone of crawfish harvesting, offering a practical and effective means of capturing these valuable crustaceans. By understanding the different types of traps, employing strategic deployment techniques, and adhering to best practices, farmers can maximize their yields while ensuring the sustainability of their operations. Addressing challenges through innovative solutions and staying abreast of technological advancements will further enhance the efficiency and environmental responsibility of crawfish farming.

5.1.2 Hand Harvesting

Hand harvesting is one of the traditional methods employed in the collection of crawfish. Despite being labor-intensive, it offers numerous advantages, including precision in selecting mature crawfish and minimizing damage to the ecosystem. This technique involves manually picking crawfish from their habitats, typically shallow ponds, rice fields, or natural waterways.

Advantages of Hand Harvesting

1. Selective Harvesting:

Hand harvesting allows for selective picking of mature crawfish, ensuring that undersized or immature crawfish are left to grow, thereby sustaining the population for future harvests.

2. Minimal Equipment:

Unlike mechanical harvesting, hand harvesting requires minimal equipment. This reduces the initial investment costs for small-scale farmers or beginners who are just starting out in crawfish farming.

3. Reduced Environmental Impact:

Hand harvesting is less disruptive to the crawfish's habitat compared to mechanical methods. It preserves the integrity of the ecosystem, which is crucial for maintaining a sustainable crawfish population.

4. Quality Control:

Harvesters can assess the quality of each crawfish during collection, ensuring that only healthy and viable crawfish are harvested. This leads to higher quality produce for market.

Disadvantages of Hand Harvesting

1. Labor-Intensive:

This method is highly labor-intensive, requiring significant human effort and time. It may not be feasible for large-scale operations where the volume of crawfish to be harvested is substantial.

2. Slower Process:

Hand harvesting is slower compared to mechanical methods, potentially limiting the amount of crawfish that can be harvested within a given timeframe.

3. Physical Demand:

The physical nature of hand harvesting can be demanding, requiring workers to bend, stoop, and often wade through water for extended periods.

Techniques and Best Practices for Hand Harvesting

To effectively harvest crawfish by hand, certain techniques and best practices should be followed:

1. Timing:

   - Crawfish are most active and easier to catch during cooler parts of the day, such as early morning or late evening. Harvesting during these times can increase efficiency.

   - Seasonal timing is also crucial. The peak harvesting season for crawfish typically ranges from late winter to early summer, depending on the region.

2. Locating Crawfish:

- Crawfish tend to hide under rocks, vegetation, and in burrows. Careful observation and understanding of their habitats are essential.

- Look for signs of crawfish activity such as burrows, muddy water, or bubbles.

3. Equipment:

- Use a small, fine-mesh net to scoop crawfish from the water. This helps in capturing them without causing harm.

- A shallow container, such as a bucket or basket, is useful for holding the harvested crawfish temporarily.

4. Handling Techniques:

- Approach crawfish slowly to avoid startling them. Quick movements can cause them to retreat into their burrows or deeper into the water.

- Grip crawfish gently but firmly behind their claws to avoid getting pinched and to minimize stress on the animal.

5. Protective Gear:

- Wearing gloves can protect your hands from sharp rocks, vegetation, and crawfish claws.

- Waterproof boots or waders are advisable to keep you dry and comfortable while working in wet environments.

Step-by-Step Process of Hand Harvesting

1. Preparation:

- Equip yourself with the necessary tools: a fine-mesh net, a bucket or basket, gloves, and waterproof boots or waders.

- Choose an optimal time of day and ensure the weather conditions are favorable for harvesting.

2. Survey the Area:

   - Walk around the pond or waterway, looking for signs of crawfish activity.

   - Identify potential hiding spots such as under rocks, in vegetation, and near burrows.

3. Capture:

   - Slowly approach the identified spots and use the net to scoop up the crawfish.

   - Carefully transfer the captured crawfish into the bucket or basket.

4. Inspection:

   - Inspect each crawfish for size and health. Return any undersized or damaged crawfish back to the water.

   - Separate the healthy, mature crawfish from those that are unfit for harvesting.

5. Post-Harvest Handling:

   - Once a sufficient quantity of crawfish has been harvested, move them to a shaded, cool area to minimize stress.

   - Follow proper cleaning and sorting procedures to prepare the crawfish for storage or sale.

Enhancing Efficiency in Hand Harvesting

While hand harvesting is inherently slower than mechanical methods, certain strategies can enhance its efficiency:

1. Team Effort:

   - Harvesting in teams can significantly speed up the process. Each member can focus on specific tasks such as locating, capturing, and inspecting crawfish.

2. Training:

   - Proper training of workers in effective harvesting techniques can improve speed and reduce the number of missed or damaged crawfish.

3. Use of Simple Tools:

   - Simple tools like long-handled nets or traps can assist in reaching difficult areas without disturbing the habitat too much.

4. Regular Monitoring:

   - Regularly monitor the crawfish population and their habitats to optimize harvesting times and ensure sustainable practices.

Conclusion

Hand harvesting remains a valuable method in the arsenal of crawfish farming techniques. Its low-cost, environmentally friendly nature makes it particularly suited for small-scale farmers and beginners. By adhering to best practices and employing strategic methods, farmers can effectively utilize hand harvesting to yield high-quality crawfish, contributing to a sustainable and profitable farming operation.

As with any agricultural practice, the key to success in hand harvesting lies in the balance between effort and reward. With careful planning, efficient techniques, and a deep understanding of crawfish behavior and habitats, hand harvesting can be an effective and rewarding method for crawfish farmers.

5.2 Post-Harvest Handling

Proper post-harvest handling of crawfish is crucial to ensure the quality and marketability of the product. This section will discuss the steps and considerations involved in cleaning and sorting crawfish after they have been harvested. Effective post-harvest handling not only maintains the freshness and quality of the crawfish but also minimizes losses due to spoilage or damage.

5.2.1 Cleaning and Sorting

Cleaning Crawfish

The first step in post-harvest handling is cleaning the crawfish to remove any dirt, debris, and other unwanted materials. Crawfish are typically harvested from ponds or natural water bodies, which means they often come into contact with mud, algae, and other organic matter. Cleaning is essential to ensure that the crawfish are safe for consumption and to enhance their appearance.

1. Initial Rinse:

   - Once the crawfish are harvested, they should be given an initial rinse with clean water to remove the bulk of the mud and debris. This can be done using a hose or by immersing the crawfish in large tanks or basins filled with water.

   - It's important to use clean, freshwater for this initial rinse to avoid introducing any contaminants that could affect the quality of the crawfish.

2. Soaking:

   - After the initial rinse, the crawfish are typically soaked in clean water for a period of time. This soaking process helps to further remove any residual dirt and allows the crawfish to purge themselves of any remaining mud or debris in their digestive systems.

- The duration of soaking can vary, but it generally lasts for several hours. In some cases, a mild salt solution may be used to encourage the purging process.

3. Second Rinse:

   - Following the soaking period, the crawfish should be given a second rinse with clean water to remove any loosened dirt and debris.

   - This rinse ensures that the crawfish are as clean as possible before moving on to the sorting stage.

4. Inspection and Manual Cleaning:

   - After the second rinse, the crawfish should be inspected for any remaining dirt or debris. Any crawfish that still have significant amounts of mud or algae attached should be manually cleaned using a brush or other suitable tool.

   - It's also important to check for and remove any dead or damaged crawfish, as these can affect the overall quality and safety of the final product.

Sorting Crawfish

Once the crawfish have been thoroughly cleaned, the next step is sorting them based on size, quality, and marketability. Sorting is a critical process that ensures uniformity in the product and helps to maximize the value of the harvest.

1. Size Sorting:

   - Crawfish are typically sorted by size using mechanical graders or by hand. Mechanical graders consist of a series of sieves or rollers with varying gaps, allowing crawfish of different sizes to be separated as they pass through the system.

   - Hand sorting involves workers manually separating the crawfish based on visual assessment of their size. This method can be more time-consuming but allows for greater precision and quality control.

2. Quality Assessment:

- In addition to size, the quality of each crawfish should be assessed during the sorting process. Quality factors to consider include the overall health and appearance of the crawfish, the absence of any visible diseases or deformities, and the presence of any damage or injuries.

- Crawfish that meet the highest quality standards can be classified as premium grade, while those with minor defects may be classified as lower grades suitable for different markets or processing.

3. Segregation of Soft-Shell Crawfish:

- Soft-shell crawfish, which have recently molted and have a softer exoskeleton, should be segregated from hard-shell crawfish. Soft-shell crawfish are more delicate and prone to damage, requiring different handling and storage procedures.

- These soft-shell crawfish are often sold at a premium due to their tender texture and are sought after for certain culinary applications.

4. Counting and Batching:

- Once sorted by size and quality, the crawfish are typically counted and batched for packaging and transportation. Counting can be done manually or using automated counting systems that ensure accurate quantities.

- Batching involves grouping the crawfish into uniform quantities, which are then placed in containers or crates for storage and transport. Proper batching helps to streamline the packaging process and ensures that each package contains a consistent quantity of crawfish.

5. Labeling and Documentation:

- Each batch of sorted crawfish should be labeled with relevant information, including the size grade, quality grade, and the date of harvest. This labeling helps with inventory management and traceability, ensuring that customers receive fresh and high-quality products.

- Maintaining detailed records of the sorting process and the characteristics of each batch is also important for quality control and regulatory compliance. These records can help to identify any issues in the handling process and provide valuable information for improving future harvests.

Best Practices for Cleaning and Sorting

To ensure the highest quality and efficiency in the cleaning and sorting process, it is important to follow best practices and implement effective management strategies. Some key best practices include:

1. Regular Equipment Maintenance:

   - Cleaning and sorting equipment should be regularly inspected and maintained to ensure optimal performance. This includes checking for any wear and tear, cleaning the equipment thoroughly after each use, and making any necessary repairs or replacements.

   - Properly maintained equipment helps to reduce the risk of contamination and ensures that the cleaning and sorting processes are as efficient and effective as possible.

2. Training and Supervision:

   - Workers involved in the cleaning and sorting process should be properly trained in the correct procedures and safety protocols. This includes understanding how to handle the crawfish gently to avoid damage, how to use the equipment correctly, and how to identify and sort crawfish based on size and quality.

   - Supervisors should monitor the process to ensure that workers are following the established procedures and to address any issues that arise promptly.

3. Hygiene and Sanitation:

- Maintaining high standards of hygiene and sanitation is crucial in the cleaning and sorting process. Workers should follow strict hygiene protocols, including wearing clean protective clothing, washing hands regularly, and using sanitized tools and equipment.

- The cleaning and sorting area should be kept clean and free of any contaminants. This includes regularly cleaning surfaces, disposing of waste properly, and using disinfectants as needed.

4. Water Quality Management:

- The quality of the water used in the cleaning and soaking processes is critical to the overall quality of the crawfish. Water should be clean, free of contaminants, and regularly tested to ensure it meets safety standards.

- If using a salt solution for soaking, the concentration should be carefully controlled to avoid causing stress or harm to the crawfish.

5. Temperature Control:

- Maintaining the proper temperature during the cleaning and sorting process is important to prevent spoilage and maintain the freshness of the crawfish. The cleaning water should be cool but not too cold, as extreme temperatures can cause stress to the crawfish.

- After sorting, the crawfish should be stored at appropriate temperatures to keep them fresh until they are transported or processed.

By following these best practices and paying careful attention to each step of the cleaning and sorting process, crawfish farmers can ensure that their harvest is of the highest quality and ready for the market. Proper post-harvest handling not only enhances the marketability of the crawfish but also contributes to the overall success and sustainability of the farming operation.

5.2.2 Storage and Transportation

Proper storage and transportation of crawfish are critical to maintaining their quality and ensuring they reach the market in prime condition. This section will provide a comprehensive guide to the best practices for storing and transporting crawfish post-harvest.

Storage

1. Immediate Post-Harvest Cooling

After harvesting, the first step is to cool the crawfish rapidly. This process helps to slow down their metabolism and reduce stress, which is crucial for maintaining their quality. Here are the recommended steps:

- Cooling Tanks: Transfer the harvested crawfish into cooling tanks filled with clean, cold water. The water temperature should be between 50°F to 60°F (10°C to 15°C). This initial cooling process should last for about 20 to 30 minutes.

- Avoiding Thermal Shock: It's important to avoid rapid temperature changes that can cause thermal shock. Gradually cool the crawfish by slowly introducing them to colder water.

2. Cleaning and Sorting

Before storage, ensure that the crawfish are properly cleaned and sorted. This step is detailed in section 5.2.1 but is crucial for preventing contamination and ensuring that only healthy, high-quality crawfish are stored.

- Rinsing: Use clean, running water to rinse the crawfish thoroughly.

- Sorting: Remove any dead, damaged, or diseased crawfish. Sorting by size and weight can also help in managing storage and transportation more efficiently.

3. Temporary Storage Before Transportation

If immediate transportation is not possible, store the crawfish in optimal conditions to preserve their quality.

- Storage Containers: Use well-ventilated plastic crates or baskets to store the crawfish. Ensure the containers allow for adequate airflow to prevent suffocation.
- Cool Environment: Store the containers in a cool, shaded area. If a cool room is available, maintain a temperature between 45°F to 55°F (7°C to 13°C). Avoid freezing temperatures as they can kill the crawfish.

4. Humidity Control

Maintaining the right humidity levels is vital for keeping crawfish alive during storage. Crawfish are highly sensitive to dehydration, and excessive moisture loss can lead to mortality.

- Moist Environment: Keep the storage environment moist. Use damp burlap sacks or towels to cover the containers. This practice helps retain humidity around the crawfish without making them too wet.
- Avoid Standing Water: Ensure that the crawfish are not sitting in standing water as this can lead to suffocation and increase the risk of bacterial growth.

Transportation

1. Preparing for Transport

Proper preparation for transportation ensures that crawfish arrive at their destination in the best possible condition.

- Pre-transport Check: Verify the health and condition of the crawfish before loading them for transport. Only transport healthy, active crawfish.
- Packaging: Pack the crawfish in well-ventilated containers. Use containers that are sturdy enough to withstand the rigors of transport.

2. Temperature Management

Maintaining the correct temperature during transportation is crucial. Crawfish are cold-blooded animals and can be severely affected by temperature fluctuations.

- Insulated Containers: Use insulated containers or coolers for transporting crawfish. These help maintain a stable temperature during transit.
- Cooling Packs: Place cooling packs or ice packs around the containers. Ensure that the ice packs are wrapped to avoid direct contact with the crawfish, which can cause thermal shock.
- Avoid Direct Sunlight: Transport crawfish in a shaded vehicle or cover the containers to avoid direct exposure to sunlight.

3. Ventilation

Adequate ventilation is essential to prevent the build-up of carbon dioxide and to ensure a steady supply of oxygen.

- Ventilated Containers: Use containers with plenty of ventilation holes. This allows for air circulation and prevents the crawfish from suffocating.

- Avoid Overcrowding: Do not overcrowd the containers. Overcrowding can lead to reduced oxygen levels and increased stress among the crawfish.

4. Duration of Transport

The length of time crawfish are in transit can significantly impact their survival and quality.

- Minimize Transport Time: Aim to minimize the duration of transport as much as possible. Shorter transport times reduce stress and mortality rates.
- Rest Stops: For longer journeys, consider rest stops to check on the crawfish and ensure they are in good condition. This is especially important if the transport time exceeds 8-10 hours.

5. Handling During Transport

Gentle handling during loading and unloading is critical to avoid injury and stress to the crawfish.

- Gentle Loading: Carefully load the containers into the transport vehicle. Avoid dropping or shaking the containers.
- Secure Placement: Ensure that the containers are securely placed in the vehicle to prevent movement during transit, which can cause stress or injury.

6. Post-Transport Handling

Upon arrival at the destination, immediate attention to the crawfish is necessary to ensure their continued quality.

- Immediate Unloading: Unload the crawfish as soon as possible. Prolonged time in the transport vehicle can increase stress and mortality.

- Recovery Period: Allow the crawfish a recovery period in a cool, well-ventilated area. If possible, place them in cool, clean water for 10-15 minutes to help them recover from the stress of transport.

- Quality Check: Perform a quality check to remove any dead or weakened crawfish. This step ensures that only the healthiest crawfish proceed to market or further storage.

Best Practices and Tips

- Regular Monitoring: During both storage and transportation, regularly monitor the temperature, humidity, and condition of the crawfish. Quick responses to any issues can prevent significant losses.

- Training for Handlers: Ensure that all personnel involved in the storage and transportation process are well-trained in handling crawfish. Proper training can significantly reduce stress and mortality rates.

- Contingency Plans: Have contingency plans in place for potential issues such as delays, equipment failure, or extreme weather conditions. Being prepared can mitigate the impact of unforeseen circumstances on the crawfish.

Conclusion

Effective storage and transportation are vital components of crawfish farming, directly impacting the profitability and sustainability of the operation. By following the guidelines outlined in this section, crawfish farmers can ensure that their product remains in top condition from harvest to market. Maintaining the right temperature, humidity, and handling practices will not only preserve the quality of the crawfish but also enhance the reputation and reliability of the farming operation.

PART VI
Marketing and Selling Crawfish

6.1 Identifying Markets

6.1.1 Local Markets

Identifying and targeting local markets is crucial for the success of any crawfish farming venture. Local markets offer a range of opportunities that are often more accessible and cost-effective compared to distant or international markets. Understanding the dynamics of local demand, consumer preferences, and the competitive landscape can help in crafting effective strategies to penetrate and dominate local markets. This section explores various aspects of local markets for crawfish, including potential customer segments, distribution channels, marketing tactics, and the benefits of focusing on local sales.

Understanding Local Demand

The first step in identifying local markets is to understand the demand for crawfish in your immediate vicinity. Conducting market research can help you gauge the level of interest and consumption patterns of local consumers. This research can involve surveys, focus groups, and analysis of sales data from local seafood markets and restaurants. Factors to consider include:

- Consumer Preferences: What types of crawfish products do local consumers prefer? Live crawfish, boiled crawfish, or processed products such as tails or étouffée?

- Seasonality: When is the peak season for crawfish consumption in your area? Are there cultural or local events that boost demand?

- Price Sensitivity: How much are local consumers willing to pay for crawfish? Understanding price elasticity can help in setting competitive prices.

Customer Segments

Identifying different customer segments within the local market is essential for targeted marketing and sales strategies. Key customer segments for crawfish include:

- Retail Consumers: These are individual buyers who purchase crawfish for personal or family consumption. Retail consumers can be targeted through direct sales at farmers' markets, roadside stands, or online platforms.

- Restaurants and Food Establishments: Local restaurants, seafood shacks, and catering companies are significant buyers of crawfish. Building relationships with these businesses can lead to steady, bulk orders.

- Grocery Stores and Supermarkets: Partnering with local grocery stores to supply fresh or frozen crawfish can expand your reach to a broader customer base.

- Event Organizers: Crawfish boils are popular at local festivals, community events, and private parties. Event organizers often need large quantities of crawfish and can become valuable repeat customers.

Distribution Channels

Effectively distributing crawfish to local markets requires a well-planned logistics strategy. Various distribution channels can be leveraged to reach different customer segments:

- Direct Sales: Selling directly to consumers through farmers' markets, roadside stands, or your own storefront can provide higher profit margins and direct customer feedback.

- Wholesale: Supplying to local restaurants, grocery stores, and seafood markets can ensure consistent sales volume. Establishing wholesale agreements with these businesses can provide a stable income stream.

- Online Sales: With the increasing trend of e-commerce, setting up an online store to sell and deliver crawfish locally can tap into a tech-savvy customer base. Offering home delivery services can add convenience for your customers.

- Subscription Services: Implementing a subscription model where customers receive regular deliveries of crawfish can create a loyal customer base and predictable revenue.

Marketing Tactics

Effective marketing is vital to attract and retain local customers. Tailored marketing tactics can help in creating awareness, generating interest, and encouraging purchases. Some proven strategies include:

- Local Advertising: Utilize local newspapers, radio stations, and community bulletin boards to advertise your crawfish products. Sponsoring local events or sports teams can also enhance brand visibility.

- Social Media: Platforms like Facebook, Instagram, and Twitter are powerful tools for reaching local audiences. Regularly posting engaging content, such as cooking tips, customer testimonials, and behind-the-scenes looks at your farming operations, can build a strong online presence.

- Community Engagement: Participating in or hosting local events such as crawfish boils, cooking competitions, or farm tours can help in building community ties and promoting your products.

- Loyalty Programs: Implementing a loyalty program where customers earn points or discounts for repeat purchases can incentivize frequent buying and foster customer loyalty.

Benefits of Focusing on Local Markets

Focusing on local markets for selling crawfish offers several advantages:

- Lower Transportation Costs: Selling locally reduces transportation expenses and the risks associated with long-distance shipping, such as spoilage and delays.

- Stronger Customer Relationships: Being close to your customers allows for better communication, personalized service, and stronger relationships. This can lead to higher customer satisfaction and repeat business.

- Community Support: Local businesses often receive strong support from their communities. By sourcing and selling locally, you contribute to the local economy, which can enhance your reputation and customer loyalty.

- Flexibility and Responsiveness: Operating within a local market allows you to quickly adapt to changes in demand, preferences, and market conditions. This agility can provide a competitive edge over larger, less nimble competitors.

Challenges and Solutions

While local markets present numerous opportunities, they also come with their own set of challenges:

- Limited Market Size: The size of the local market may be limited, which could restrict the growth potential. To mitigate this, continually seek new customer segments and consider gradual expansion to neighboring areas.

- Competition: Competing with established local suppliers can be tough. Differentiating your products through quality, unique offerings, or superior customer service can help you stand out.

- Regulatory Compliance: Navigating local regulations and health standards is crucial. Staying informed about local laws and maintaining high standards of hygiene and product quality can prevent legal issues and build customer trust.

- Market Saturation: In areas with a high number of crawfish suppliers, the market can become saturated. Regularly innovating your product offerings and marketing strategies can keep your business competitive.

Case Studies

Examining successful case studies of crawfish farmers who have thrived in local markets can provide valuable insights and inspiration:

- Example 1: Joe's Crawfish Farm: Located in a small town, Joe's Crawfish Farm leveraged local festivals and community events to promote their products. By sponsoring a popular annual crawfish boil event, they significantly boosted their brand recognition and sales.
- Example 2: Bayou Fresh Crawfish: This farm focused on direct sales through a weekly farmers' market. They built a loyal customer base by offering cooking demonstrations and free samples, which helped to educate consumers and create demand.
- Example 3: Riverside Crawfish: Partnering with local restaurants, Riverside Crawfish ensured a steady demand for their products. They offered special pricing for bulk orders and collaborated on menu promotions, which increased their market penetration.

In conclusion, identifying and targeting local markets for crawfish can provide a solid foundation for your business. By understanding local demand, segmenting customers, leveraging effective distribution channels, and employing targeted marketing tactics, you can establish a strong presence in your community. The benefits of focusing on local markets, combined with the ability to build close customer relationships and support the local economy, make this a worthwhile strategy for any crawfish farmer.

6.1.2 Export Opportunities

Exporting crawfish can be a lucrative venture for farmers looking to expand their market reach beyond local and regional boundaries. The global demand for crawfish is rising, driven by its growing popularity as a delicacy in various cuisines and its increasing use in gourmet dishes worldwide. This section will explore the potential export opportunities for crawfish farmers, focusing on market research, regulatory considerations, logistics, and strategies for successful entry into international markets.

Market Research and Analysis

The first step in identifying export opportunities is thorough market research. Understanding which countries have a high demand for crawfish, their consumption patterns, and preferences is crucial. The primary markets for crawfish exports include the United States, China, European Union countries, and Southeast Asia. Each of these markets has unique characteristics that need to be considered.

- United States: The U.S. has a significant market for seafood, including crawfish, especially in states like Louisiana, where crawfish boils are a cultural tradition. The demand peaks during the spring and early summer. Understanding the seasonality and regional preferences within the U.S. can help target marketing efforts more effectively.

- China: China is one of the largest consumers of seafood globally. Crawfish is increasingly popular in urban areas, often served in hot pot dishes and other culinary preparations. The growing middle class with higher disposable incomes is driving this demand. Exporters should consider local partnerships to navigate the complex distribution channels in China.

- European Union: The EU, particularly countries like France and Spain, shows a growing interest in crawfish, often used in gourmet cooking. The EU market requires strict adherence to food safety and quality standards, so compliance with these regulations is crucial for successful entry.

- Southeast Asia: Countries such as Thailand and Vietnam also present growing markets for crawfish. These markets are characterized by a high demand for fresh and live seafood, and logistical considerations for maintaining product quality during transport are vital.

Regulatory Considerations

Navigating the regulatory landscape is a critical aspect of exporting crawfish. Each country has its own set of regulations governing the import of seafood, including crawfish. These regulations often cover food safety, quality standards, packaging, labeling, and documentation requirements.

- Food Safety and Quality Standards: Exporters must ensure that their products meet the food safety standards of the destination country. This includes adherence to HACCP (Hazard Analysis and Critical Control Points) principles, traceability, and regular inspections. For instance, the EU has stringent requirements under the General Food Law Regulation.

- Packaging and Labeling: Proper packaging is essential to maintain the quality and freshness of crawfish during transportation. Packaging should be durable, temperature-controlled, and compliant with the importing country's labeling requirements, which may include details about the product's origin, processing date, and nutritional information.

- Documentation: Accurate and complete documentation is necessary for smooth customs clearance. This includes health certificates, certificates of origin, and export licenses. Working with a customs broker can help ensure that all paperwork is correctly filled out and submitted.

Logistics and Supply Chain Management

Effective logistics and supply chain management are pivotal in ensuring that crawfish reach international markets in prime condition. The perishable nature of seafood requires careful planning and execution of the supply chain processes.

- Cold Chain Management: Maintaining the cold chain from harvest to delivery is essential to preserve the quality of crawfish. This involves using refrigerated trucks, cold storage facilities, and temperature-controlled shipping containers. Any break in the cold chain can result in spoilage and loss of product.

- Shipping Methods: Depending on the destination, crawfish can be transported by air or sea. Air freight is faster and suitable for live crawfish or markets that require rapid delivery, while sea freight is more cost-effective for frozen products. The choice of shipping method should balance cost, speed, and the nature of the product.

- Distribution Channels: Establishing reliable distribution channels in the destination country is crucial. This can involve partnerships with local distributors, retailers, or wholesalers who have established networks and market knowledge. Such partnerships can help in navigating local market dynamics and consumer preferences.

Strategies for Successful Market Entry

To successfully enter and establish a presence in international markets, crawfish exporters should adopt strategic approaches tailored to each market's unique characteristics.

- Market Entry Modes: Exporters can choose from various market entry modes, such as direct exporting, joint ventures, or partnerships with local firms. Direct exporting involves selling directly to customers in the foreign market, offering more control over the brand and pricing. Joint ventures or partnerships can provide local market insights and reduce entry barriers.

- Branding and Promotion: Building a strong brand is essential for differentiation in competitive markets. Exporters should invest in creating a compelling brand story that highlights the quality, origin, and unique attributes of their crawfish. Promotional activities can include participation in international seafood trade shows, digital marketing campaigns, and collaborations with local chefs and restaurants.

- Cultural Adaptation: Understanding and adapting to the cultural preferences and culinary practices of the target market can enhance product acceptance. For instance, packaging sizes, cooking instructions, and flavor profiles can be adjusted to suit local tastes.

- Customer Relationships: Building and maintaining strong relationships with customers and partners in the export market is crucial. Providing excellent customer service, timely delivery, and consistent product quality can help establish long-term business relationships.

Case Study: Exporting Crawfish to China

A practical example of successful export strategies can be seen in the case of crawfish exports to China. Several crawfish farmers in the United States have tapped into the Chinese market by adopting a combination of direct exporting and partnerships with local distributors. They have invested in state-of-the-art cold chain logistics to ensure that live crawfish arrive fresh and in optimal condition.

Marketing campaigns in China have focused on educating consumers about the versatility of crawfish in various dishes and promoting it as a premium product. Collaborations with popular Chinese food bloggers and chefs have also helped in creating buzz and increasing demand.

To comply with Chinese regulations, these exporters have ensured that their products meet the required food safety standards and have obtained the necessary certifications. Additionally, they have tailored their packaging to include Mandarin labels with detailed product information, which has enhanced consumer trust and product acceptance.

Challenges and Mitigation Strategies

Despite the opportunities, exporting crawfish also comes with challenges. These include fluctuating international demand, currency exchange risks, and potential trade barriers. Exporters must be prepared to address these challenges proactively.

- Demand Fluctuations: International markets can be volatile, with demand influenced by factors such as economic conditions, consumer trends, and seasonal variations. Exporters should diversify their markets to reduce dependence on any single market and engage in continuous market research to anticipate changes.

- Currency Risks: Exchange rate fluctuations can impact profitability. Exporters can use hedging strategies to mitigate these risks, such as forward contracts or options to lock in favorable exchange rates.

- Trade Barriers: Tariffs, quotas, and other trade barriers can affect market access. Staying informed about trade policies and engaging in advocacy through industry associations can help address these issues. Additionally, seeking support from government export promotion programs can provide valuable resources and assistance.

Conclusion

Exporting crawfish presents significant growth opportunities for farmers willing to navigate the complexities of international trade. By conducting thorough market research, complying with regulatory requirements, ensuring efficient logistics, and adopting strategic market entry approaches, exporters can successfully tap into the global demand for crawfish. Building strong customer relationships and continuously adapting to market changes will be key to sustaining long-term success in the export market.

6.2 Pricing Strategies

Effective pricing strategies are critical to the success of any crawfish farming operation. This section delves into the methodologies and considerations involved in setting prices for your crawfish, ensuring profitability while remaining competitive in the market. By understanding cost analysis, you can make informed pricing decisions that reflect the true value of your product and sustain your business over the long term.

6.2.1 Cost Analysis

Cost analysis is the backbone of any pricing strategy. It involves a thorough examination of all costs associated with the production and sale of crawfish. This analysis helps farmers determine the minimum price at which they can sell their crawfish without incurring losses. The following components are essential in conducting a comprehensive cost analysis for your crawfish farming operation:

1. Direct Costs

Direct costs are expenses that can be directly attributed to the production of crawfish. These include:

- Seed Crawfish: The cost of purchasing seed crawfish or juvenile crawfish that will be grown to market size. This cost varies depending on the source and quality of the crawfish.

- Feed: Crawfish require a balanced diet to grow efficiently. The cost of feed can fluctuate based on the type and source of the feed, as well as market conditions.

- Labor: Wages paid to workers involved in the daily operations of the farm, including feeding, monitoring, and harvesting the crawfish.

- Supplies and Equipment: Costs for nets, traps, tanks, aerators, and other equipment necessary for maintaining and harvesting crawfish.

2. Indirect Costs

Indirect costs are expenses that are not directly tied to the production process but are necessary for the overall operation. These include:

- Utilities: Electricity, water, and other utility costs for maintaining the appropriate environment for crawfish growth.

- Maintenance and Repairs: Regular upkeep of equipment and infrastructure to ensure smooth operation and prevent costly breakdowns.

- Depreciation: The gradual reduction in value of farm equipment and infrastructure over time, which should be accounted for in pricing.

3. Fixed Costs

Fixed costs remain constant regardless of the level of production. These include:

- Lease or Mortgage Payments: Costs associated with the land or facilities used for farming.

- Insurance: Coverage for property, equipment, and liability, protecting against potential risks and losses.

- Licenses and Permits: Fees for obtaining the necessary legal permissions to operate a crawfish farm.

4. Variable Costs

Variable costs change in direct proportion to the volume of production. These include:

- Transportation: Costs for delivering crawfish to markets or customers. This can include fuel, vehicle maintenance, and driver wages.
- Packaging: Materials and labor involved in packaging crawfish for sale, ensuring they are kept fresh and presented attractively.

5. Overhead Costs

Overhead costs encompass the general expenses required to run the business but not directly tied to production. These include:

- Administrative Expenses: Salaries for management and administrative staff, office supplies, and other related costs.
- Marketing and Sales: Expenses related to advertising, promotional materials, and sales commissions.

6. Contingency Costs

Contingency costs are funds set aside to cover unexpected expenses or emergencies. This could include:

- Disease Outbreaks: Costs associated with treating and managing diseases that may affect the crawfish.

- Natural Disasters: Funds to repair damage caused by floods, storms, or other natural events.

7. Break-Even Analysis

A break-even analysis helps determine the minimum amount of crawfish that must be sold at a given price to cover all costs. This analysis is crucial for setting a baseline price and understanding the volume of sales needed to achieve profitability. The formula for break-even analysis is:

$$\text{Break-Even Point (units)} = \frac{\text{Total Fixed Costs}}{\text{Price per Unit} - \text{Variable Cost per Unit}}$$

This calculation helps farmers identify the sales target required to cover their costs and start generating profit.

8. Profit Margin

Setting a price also involves deciding on the desired profit margin. The profit margin is the percentage of profit over the cost price, ensuring that the business is sustainable and can invest in growth and development. The formula for calculating profit margin is:

$$\text{Profit Margin (\%)} = \left(\frac{\text{Selling Price} - \text{Cost Price}}{\text{Selling Price}} \right) \times 100$$

A realistic and competitive profit margin should be chosen based on market conditions and business goals.

9. Market Conditions

Understanding market conditions is vital for effective pricing. This involves:

- Demand and Supply: Assessing the demand for crawfish in your target markets and the supply provided by competitors. High demand with limited supply can justify higher prices.

- Seasonality: Recognizing the seasonal fluctuations in crawfish availability and demand, and adjusting prices accordingly to optimize sales throughout the year.

- Competitor Pricing: Analyzing the pricing strategies of competitors to ensure your prices are competitive while still covering costs and achieving desired profit margins.

10. Value Proposition

Your pricing should reflect the value proposition of your crawfish. Factors that can enhance the perceived value include:

- Quality: Superior quality crawfish can command higher prices.

- Sustainability: Environmentally friendly and sustainable farming practices can appeal to eco-conscious consumers, allowing for premium pricing.

- Brand Reputation: A strong, trusted brand can justify higher prices through customer loyalty and perceived value.

11. Dynamic Pricing Strategies

Consider implementing dynamic pricing strategies to optimize revenue. These can include:

- Promotional Pricing: Offering discounts or special deals during off-peak seasons or to stimulate sales.

- Tiered Pricing: Different pricing tiers based on the size or grade of crawfish, catering to diverse customer segments.

- Subscription Models: Providing regular deliveries of crawfish at a discounted rate for long-term customers.

12. Monitoring and Adjusting Prices

Regularly review and adjust your prices based on cost changes, market conditions, and sales performance. Tools such as financial software and market analysis reports can assist in tracking these variables and making informed pricing decisions.

Conclusion

Conducting a thorough cost analysis is crucial for developing effective pricing strategies for your crawfish farming business. By understanding and accounting for all direct, indirect, fixed, variable, overhead, and contingency costs, you can set prices that ensure profitability and competitiveness. Regularly monitoring market conditions and adjusting prices as needed will help maintain a sustainable and successful business.

6.2.2 Competitive Pricing

Determining the right pricing strategy is crucial for the success of any crawfish farming business. It requires a deep understanding of various cost factors, market dynamics, and competitive landscape. This section explores different pricing strategies, focusing on competitive pricing to ensure your crawfish products are attractively priced to attract customers while ensuring profitability.

Competitive pricing is a strategy where the price of your product is set based on the prices of similar products offered by competitors. This method is particularly important in markets where there are many producers and price competition is fierce. The goal is to set a price that is competitive but also covers your costs and provides a reasonable profit margin.

Understanding the Market

The first step in competitive pricing is understanding the market. You need to identify who your competitors are and what prices they are charging. This can be done through market research, which involves gathering data on competitors' prices, their product offerings, and their market positioning. Key sources of information include:

- Competitor Websites and Online Marketplaces: Review prices listed on competitors' websites and major online marketplaces.

- Market Reports and Industry Publications: These often contain comprehensive pricing data and market analyses.

- Surveys and Customer Feedback: Direct feedback from customers can provide insights into price sensitivity and preferences.

Analyzing Competitor Pricing

Once you have gathered data, analyze it to identify pricing trends and patterns. Look for the following:

- Price Ranges: Identify the highest, lowest, and average prices in the market.

- Product Differentiation: Understand how competitors justify their pricing. This could be based on product quality, brand reputation, or additional services.

- Market Segments: Determine if there are different price points for different market segments, such as wholesale vs. retail or domestic vs. export markets.

Setting Your Price

With a clear understanding of the competitive landscape, you can set your price. Here are some approaches to consider:

1. Match Competitor Prices: This is a straightforward approach where you set your price at the same level as your main competitors. This strategy can be effective if you believe your product offers similar value and you want to compete directly.

2. Price Slightly Lower: Setting your price slightly lower than competitors can attract price-sensitive customers. However, this approach requires careful cost management to ensure profitability.

3. Price Slightly Higher: If your product offers additional value, such as higher quality, better service, or unique features, you can justify a higher price. This strategy relies on convincing customers that the added value is worth the extra cost.

Cost Considerations

Regardless of the pricing strategy, it's essential to ensure that the price covers all costs and provides a profit margin. Key cost components to consider include:

- Production Costs: This includes the cost of feed, labor, equipment, and other inputs.

- Overhead Costs: Fixed costs such as rent, utilities, and administrative expenses.

- Distribution Costs: Costs associated with transporting the product to market, including packaging, shipping, and handling.

- Marketing Costs: Expenses related to promoting the product and building brand awareness.

Conduct a thorough cost analysis to determine the minimum price you need to cover these costs. This will help you avoid setting a price that is too low to sustain your business.

Dynamic Pricing

Competitive pricing is not a one-time activity. The market is constantly changing, and so should your pricing strategy. Dynamic pricing involves regularly reviewing and adjusting prices based on market conditions, competitor actions, and changes in costs. This can be achieved through:

- Regular Market Monitoring: Continuously track competitor prices and market trends.

- Adjusting for Seasonality: Crawfish demand can vary seasonally, and prices may need to be adjusted accordingly.

- Promotional Pricing: Temporary price reductions or special offers can attract new customers and boost sales during slow periods.

Psychological Pricing

In addition to the competitive analysis, consider psychological pricing tactics that can influence customer perception and behavior. Examples include:

- Charm Pricing: Setting prices just below a round number (e.g., $9.99 instead of $10.00) can make the price appear more attractive.

- Bundling: Offering a bundle of products at a slightly lower price than the total of individual items can encourage customers to buy more.

- Price Anchoring: Presenting a higher-priced item alongside a standard option can make the standard option seem more affordable.

Communicating Value

Effectively communicating the value of your product is essential when using competitive pricing. Even if your price is similar to or slightly higher than competitors, you need to convince customers that your product offers superior value. This can be achieved through:

- Clear Messaging: Highlight the unique features and benefits of your product.

- Customer Testimonials: Use positive reviews and testimonials to build trust and credibility.

- Quality Assurance: Emphasize quality control measures and certifications that assure customers of the product's quality.

Legal and Ethical Considerations

When setting prices, it's important to be aware of legal and ethical considerations. Avoid practices that could be seen as predatory pricing (setting prices extremely low to drive competitors out of the market) or price fixing (colluding with competitors to set prices). Adhering to fair competition laws helps maintain a healthy market environment and avoids legal repercussions.

Conclusion

Competitive pricing is a strategic tool that requires careful analysis and ongoing management. By understanding the market, analyzing competitor pricing, and setting a price that reflects both your costs and the competitive landscape, you can position your crawfish farming business for success. Remember that pricing is not static; it should evolve with market conditions and customer expectations. Through dynamic pricing, psychological tactics, and clear communication of value, you can attract and retain customers while ensuring your business remains profitable.

6.3 Branding and Promotion

6.3.1 Creating a Brand

Creating a strong and memorable brand is a critical component of any business, including crawfish farming. A well-established brand can differentiate your product from competitors, create customer loyalty, and increase market value. In this section, we will explore the essential steps and strategies for creating a compelling brand for your crawfish farming business.

Understanding Your Brand Identity

The first step in creating a brand is understanding your brand identity. This involves defining who you are, what you stand for, and what makes your crawfish unique. Consider the following aspects:

- Mission Statement: What is the purpose of your business? Is it to provide high-quality, sustainably farmed crawfish? Or perhaps to deliver the freshest product to your local market? Your mission statement should reflect your core values and objectives.

- Vision Statement: Where do you see your business in the future? Your vision statement should articulate your long-term goals and aspirations.

- Core Values: What principles guide your business operations? These could include sustainability, quality, customer satisfaction, and innovation.

- Unique Selling Proposition (USP): What sets your crawfish apart from the competition? This could be your farming practices, the freshness of your product, your commitment to sustainability, or your customer service.

Developing Your Brand Elements

Once you have a clear understanding of your brand identity, the next step is to develop the tangible elements that will represent your brand. These elements include:

- Brand Name: Your brand name should be memorable, easy to pronounce, and reflective of your business. It should convey the essence of your brand and resonate with your target audience.

- Logo: Your logo is a visual representation of your brand. It should be simple, versatile, and easily recognizable. Consider working with a professional designer to create a logo that embodies your brand's values and appeals to your market.

- Tagline: A tagline is a short, catchy phrase that encapsulates your brand's essence. It should be memorable and convey a key benefit or characteristic of your product.

- Brand Colors and Fonts: Choose colors and fonts that reflect your brand's personality and are consistent across all marketing materials. Colors can evoke certain emotions and associations, so select those that align with your brand message.

Crafting Your Brand Story

A compelling brand story can help you connect with your audience on an emotional level. Your brand story should include:

- The Origin of Your Business: Share how and why you started your crawfish farming business. Was it a family tradition, a passion for sustainable farming, or a desire to provide fresh, local seafood?

- Challenges and Triumphs: Highlight any significant challenges you faced and how you overcame them. This can make your brand more relatable and trustworthy.

- Customer Success Stories: Share testimonials and stories from satisfied customers. This adds credibility to your brand and showcases the value of your product.

Implementing Your Brand Strategy

With your brand identity and elements in place, it's time to implement your brand strategy. This involves consistently applying your brand across all touchpoints, including:

- Packaging: Your packaging should reflect your brand's identity and stand out on the shelves. It should be attractive, functional, and informative, providing essential details about your product and brand.

- Website: Your website is often the first point of contact for potential customers. Ensure it is well-designed, easy to navigate, and reflects your brand's identity. Include compelling content, high-quality images, and clear calls to action.

- Social Media: Utilize social media platforms to engage with your audience and promote your brand. Share your brand story, behind-the-scenes glimpses of your farm, customer testimonials, and relevant content that resonates with your audience.

- Advertising: Develop advertising campaigns that align with your brand's identity and target your desired audience. This could include online ads, print ads, radio spots, or sponsorships.

Building Brand Awareness

Building brand awareness is crucial for attracting new customers and growing your business. Strategies for building brand awareness include:

- Public Relations: Leverage PR to gain media coverage and increase visibility. This could involve pitching stories to local newspapers, magazines, and online publications, or participating in community events.

- Partnerships and Collaborations: Partner with other businesses, chefs, or influencers who align with your brand values. This can help you reach a broader audience and enhance your credibility.

- Community Engagement: Participate in local events, farmers' markets, and festivals to connect with your community and showcase your brand. Sponsoring community events or offering farm tours can also enhance your brand visibility.

- Online Presence: Invest in SEO (Search Engine Optimization) to ensure your website ranks well in search results. Regularly update your website with fresh content, and engage with your audience on social media to maintain an active online presence.

Maintaining Brand Consistency

Consistency is key to building a strong brand. Ensure that all your branding elements are consistently applied across all platforms and touchpoints. This includes:

- Visual Consistency: Use the same logo, colors, and fonts across all marketing materials, packaging, and digital platforms.

- Message Consistency: Ensure your brand message is consistent in all communications. This includes your website, social media, advertising, and customer service interactions.

- Experience Consistency: Provide a consistent customer experience across all touchpoints. This includes the quality of your product, the buying process, and post-purchase support.

Measuring Brand Success

Finally, it's important to measure the success of your branding efforts. Use the following metrics to assess your brand performance:

- Brand Awareness: Track metrics such as website traffic, social media engagement, and media coverage to gauge how well your brand is known.

- Customer Loyalty: Monitor repeat purchase rates, customer feedback, and Net Promoter Scores (NPS) to measure customer loyalty and satisfaction.

- Market Share: Analyze your sales data and market share to understand your position in the market relative to competitors.

- Brand Perception: Conduct surveys and focus groups to gather insights into how your brand is perceived by customers and stakeholders.

Adapting and Evolving Your Brand

As your business grows and market conditions change, it's important to regularly review and adapt your brand strategy. Stay attuned to customer feedback, industry trends, and competitive dynamics to ensure your brand remains relevant and compelling.

In summary, creating a strong brand for your crawfish farming business involves understanding your brand identity, developing key brand elements, crafting a compelling brand story, implementing a consistent brand strategy, building brand awareness, maintaining consistency, and measuring success. By following these steps, you can establish a brand that resonates with your audience, differentiates your product, and drives business growth.

6.3.2 Marketing Campaigns

Creating a successful marketing campaign is essential for the growth and profitability of your crawfish farming business. An effective marketing campaign helps in building brand awareness, attracting customers, and driving sales. This section will provide a comprehensive guide on how to design and execute marketing campaigns that can help your crawfish business thrive.

Understanding Your Audience

PART VI: MARKETING AND SELLING CRAWFISH

The first step in developing a marketing campaign is to understand your target audience. Knowing who your potential customers are, their preferences, behaviors, and buying patterns is crucial. Here are some key steps to understand your audience:

- Market Research: Conduct surveys, focus groups, and research to gather information about your target market. Analyze demographic data, including age, gender, income level, and geographic location.

- Customer Profiles: Create detailed customer profiles or personas that represent your ideal customers. These profiles should include information about their lifestyle, purchasing habits, and preferences.

- Competitor Analysis: Study your competitors to understand who their customers are and what marketing strategies they are using. This can provide insights into your target market and help you differentiate your brand.

Setting Clear Objectives

Once you understand your audience, the next step is to set clear marketing objectives. These objectives should be specific, measurable, achievable, relevant, and time-bound (SMART). Common objectives for a marketing campaign in the crawfish farming industry might include:

- Increasing brand awareness within a specific geographic area.

- Boosting sales during peak crawfish seasons.

- Expanding into new markets or customer segments.

- Enhancing customer engagement and loyalty.

Crafting a Compelling Message

Your marketing message should communicate the unique value proposition of your crawfish products. It should highlight what sets your crawfish apart from competitors and why customers should choose your brand. Consider the following elements when crafting your message:

- Brand Story: Share the story behind your crawfish farm, including its history, values, and mission. This helps build an emotional connection with your audience.

- Product Quality: Emphasize the quality, freshness, and sustainability of your crawfish. Highlight any unique farming practices or certifications that set your product apart.

- Customer Benefits: Focus on the benefits that customers will experience by choosing your crawfish, such as superior taste, freshness, or health benefits.

Choosing the Right Marketing Channels

Selecting the right marketing channels is critical to reaching your target audience effectively. Here are some key channels to consider for your crawfish farming business:

- Social Media: Platforms like Facebook, Instagram, and Twitter are excellent for building brand awareness and engaging with customers. Use high-quality images and videos to showcase your crawfish and share behind-the-scenes content from your farm.

- Website and Blog: A well-designed website serves as the hub of your online presence. Include detailed information about your products, farming practices, and contact details. Maintain a blog with recipes, cooking tips, and updates about your farm.

- Email Marketing: Build an email list of customers and prospects. Send regular newsletters with promotions, product updates, and useful content. Personalize your emails to make them more relevant to the recipients.

- Local Advertising: Utilize local newspapers, magazines, and radio stations to promote your crawfish. Participate in community events and sponsor local activities to increase your visibility.

- Online Marketplaces: List your crawfish on online marketplaces and food delivery platforms to reach a broader audience. Ensure your listings are attractive and provide detailed information about your products.

Implementing and Monitoring Your Campaign

With your objectives set, message crafted, and channels selected, it's time to implement your marketing campaign. Here are some key steps to ensure a successful implementation:

- Content Creation: Develop engaging and high-quality content for each marketing channel. This includes social media posts, blog articles, email newsletters, and advertisements. Use professional photography and video to showcase your crawfish.

- Scheduling and Posting: Create a content calendar to plan and schedule your marketing activities. Consistency is key, so ensure you post regularly and at optimal times for your audience.

- Engagement and Interaction: Respond to comments, messages, and reviews promptly. Engaging with your audience helps build relationships and encourages customer loyalty.

- Monitoring and Analytics: Use analytics tools to track the performance of your marketing campaigns. Monitor key metrics such as website traffic, social media engagement, email open rates, and sales conversions. Analyze the data to identify what's working and what needs improvement.

Evaluating and Adjusting Your Strategy

Marketing is an ongoing process that requires continuous evaluation and adjustment. After your campaign has run for a sufficient period, assess its effectiveness against your objectives. Here are some steps to evaluate and refine your strategy:

- Performance Review: Compare the results of your campaign with your initial objectives. Identify which goals were met and which were not. Analyze the reasons behind the success or failure of specific tactics.

- Customer Feedback: Collect feedback from your customers to understand their perceptions of your brand and marketing efforts. Use surveys, reviews, and direct interactions to gather insights.

- Adjustments and Improvements: Based on your performance review and customer feedback, make necessary adjustments to your marketing strategy. This could involve tweaking your message, changing your marketing channels, or refining your target audience.

Leveraging Seasonal Opportunities

Crawfish farming is often influenced by seasonal demand, with peak seasons providing excellent opportunities for targeted marketing campaigns. Here's how to leverage seasonal opportunities:

- Seasonal Promotions: Plan special promotions and discounts during peak crawfish seasons. Create limited-time offers that encourage customers to make a purchase.

- Holiday Marketing: Align your marketing campaigns with holidays and special events. For example, promote crawfish boils for Memorial Day, Fourth of July, or Labor Day celebrations.

- Recipe and Cooking Tips: Share seasonal recipes and cooking tips that feature your crawfish. This can inspire customers to try new dishes and increase their consumption.

Collaborations and Partnerships

Building collaborations and partnerships can amplify the reach of your marketing campaigns. Consider partnering with local businesses, chefs, influencers, and organizations that align with your brand values. Here are some ideas:

- Local Restaurants and Chefs: Collaborate with local restaurants and chefs to feature your crawfish on their menus. This not only promotes your product but also adds credibility to your brand.

- Influencer Marketing: Partner with food bloggers, chefs, and social media influencers who have a strong following in your target market. Their endorsements can help increase your brand's visibility and credibility.

- Community Events: Participate in local food festivals, farmers' markets, and community events. Set up a booth to showcase your crawfish, offer samples, and engage with potential customers.

Building Long-Term Customer Relationships

Successful marketing campaigns should focus on building long-term relationships with customers rather than just driving one-time sales. Here are some strategies to foster customer loyalty:

- Customer Loyalty Programs: Implement loyalty programs that reward repeat customers. Offer discounts, exclusive access to new products, or special promotions for loyal customers.

- Personalized Communication: Use customer data to personalize your communication. Address customers by their names, send personalized offers, and tailor your messages based on their preferences and past purchases.

- Quality Customer Service: Provide excellent customer service at all touchpoints. Respond promptly to inquiries, address complaints professionally, and go the extra mile to exceed customer expectations.

Measuring ROI

Finally, it's important to measure the return on investment (ROI) of your marketing campaigns. This helps you understand the financial impact of your marketing efforts and make informed decisions for future campaigns. Here's how to measure ROI:

- Calculate Costs: Track all costs associated with your marketing campaign, including advertising spend, content creation expenses, and any other related costs.

- Track Revenue: Measure the revenue generated from your marketing campaign. This can include direct sales, increased website traffic, and new customer acquisitions.

- Analyze Metrics: Use key performance indicators (KPIs) such as cost per acquisition (CPA), customer lifetime value (CLV), and conversion rates to evaluate the effectiveness of your campaign.

- Compare and Optimize: Compare the ROI of different marketing campaigns to identify which strategies are most effective. Use this data to optimize future campaigns and allocate your marketing budget more efficiently.

Conclusion

A well-executed marketing campaign is crucial for the success of your crawfish farming business. By understanding your audience, setting clear objectives, crafting a compelling message, choosing the right channels, and continuously evaluating your efforts, you can create effective marketing campaigns that drive growth and profitability. Remember, marketing is an ongoing process that requires creativity, consistency, and adaptability to achieve long-term success.

PART VI: MARKETING AND SELLING CRAWFISH

PART VII
Sustainable Crawfish Farming Practices

7.1 Environmental Considerations

In the realm of sustainable crawfish farming, understanding and implementing environmentally conscious practices is paramount. The key to achieving a balance between high productivity and ecological responsibility lies in adopting strategies that minimize the environmental impact of farming activities. This section explores various eco-friendly farming techniques that crawfish farmers can utilize to ensure their operations are both sustainable and profitable.

7.1.1 Eco-Friendly Farming Techniques

Eco-friendly farming techniques are designed to reduce the environmental footprint of agricultural practices while maintaining or enhancing farm productivity. In crawfish farming, these techniques involve careful planning and execution of various methods that promote ecological balance, conserve resources, and minimize pollution. Here, we delve into several effective eco-friendly techniques specifically tailored for crawfish farming.

1. Integrated Pest Management (IPM):

Integrated Pest Management (IPM) is a holistic approach to pest control that reduces reliance on chemical pesticides, which can have harmful environmental effects. IPM in crawfish farming involves:

- Monitoring Pest Populations: Regularly inspect crawfish ponds for pests and their activity. This helps in early detection and identification, allowing for targeted interventions.

- Biological Controls: Introduce natural predators or beneficial organisms that can help control pest populations. For example, fish species that do not compete with crawfish can be introduced to control mosquito larvae.

- Cultural Practices: Modify farming practices to reduce pest habitats. For instance, managing water levels can disrupt the breeding cycles of pests.

- Mechanical Controls: Utilize traps or barriers to physically remove or prevent pests from accessing the crawfish.

By combining these methods, farmers can significantly reduce the need for chemical pesticides, thereby protecting the environment and promoting a healthier ecosystem.

2. Water Management:

Efficient water management is crucial in crawfish farming, as it directly impacts both the health of the crawfish and the surrounding environment. Key water management practices include:

- Water Recycling: Implementing systems that allow for the recycling and reuse of water within the farm. This reduces the need for fresh water and minimizes wastewater discharge.

- Wetland Conservation: Protecting and restoring natural wetlands around the farm. Wetlands act as natural filters, improving water quality by trapping sediments and pollutants.

- Optimal Water Use: Monitoring and adjusting water levels to ensure they are optimal for crawfish growth while avoiding excessive water usage. This can be achieved through automated water level sensors and controls.

Effective water management not only conserves a vital resource but also ensures that water leaving the farm is cleaner, reducing the impact on local water bodies.

3. Soil Health and Management:

Healthy soil is fundamental to sustainable farming. For crawfish farming, maintaining soil health involves:

- Cover Cropping: Planting cover crops during the off-season to prevent soil erosion, improve soil structure, and add organic matter to the soil. Leguminous cover crops can also fix nitrogen, enhancing soil fertility.

- Minimal Tillage: Reducing tillage to maintain soil structure and prevent erosion. Minimal tillage practices help retain soil moisture and organic matter, which are beneficial for crawfish habitats.

- Organic Amendments: Using organic fertilizers and compost to enrich the soil with nutrients. Organic amendments improve soil health over time, reducing the need for chemical fertilizers.

By focusing on soil health, farmers can create a more resilient farming system that supports sustainable crawfish production.

4. Habitat Enhancement:

Creating and maintaining diverse habitats within and around crawfish ponds can enhance biodiversity and ecological balance. Habitat enhancement practices include:

- Vegetative Buffers: Planting vegetation along pond edges to serve as buffers that trap sediments, nutrients, and pollutants before they enter the water. These buffers also provide habitats for beneficial insects and wildlife.

- Aquatic Vegetation: Encouraging the growth of beneficial aquatic plants within ponds. These plants provide shelter and food for crawfish and other aquatic organisms, promoting a balanced ecosystem.

- Biodiversity Corridors: Establishing corridors of natural vegetation that connect different habitats. These corridors facilitate the movement of wildlife and promote biodiversity.

Habitat enhancement not only benefits crawfish but also supports a wide range of wildlife, contributing to the overall health of the ecosystem.

5. Energy Efficiency:

Reducing energy consumption is an important aspect of eco-friendly farming. Strategies to improve energy efficiency in crawfish farming include:

- Renewable Energy Sources: Installing solar panels or wind turbines to generate renewable energy for farm operations. This reduces reliance on fossil fuels and lowers greenhouse gas emissions.

- Energy-Efficient Equipment: Using energy-efficient pumps, aerators, and other equipment. Regular maintenance and upgrades to more efficient models can significantly reduce energy usage.

- Smart Technology: Implementing smart technology and automation to optimize energy use. For example, automated systems can adjust aeration and water flow based on real-time data, ensuring energy is used only when necessary.

By adopting energy-efficient practices, farmers can reduce their carbon footprint and operational costs.

6. Waste Management:

Effective waste management is crucial to minimizing the environmental impact of crawfish farming. Practices include:

- Waste Reduction: Implementing strategies to reduce waste generation at the source. This can involve optimizing feed use to minimize leftovers and improving harvesting techniques to reduce bycatch.

- Recycling and Reuse: Recycling waste materials whenever possible. For example, shells and other organic waste can be composted and used as fertilizer.

- Proper Disposal: Ensuring that any waste that cannot be reused or recycled is disposed of properly. This includes following regulations for the disposal of hazardous materials and using approved disposal sites.

Proper waste management helps prevent pollution and promotes a cleaner, healthier environment.

7. Sustainable Feed Practices:

The type and amount of feed used in crawfish farming can significantly impact the environment. Sustainable feed practices include:

- Natural Feeding: Maximizing the use of natural food sources available in the ponds, such as algae and detritus. This reduces the need for supplemental feed.

- Eco-Friendly Feed: Using feed that is formulated from sustainable sources and has minimal environmental impact. This includes selecting feed with high digestibility to reduce waste.

- Efficient Feeding Techniques: Implementing feeding techniques that minimize feed wastage, such as timed feeding schedules and using feeding trays.

By focusing on sustainable feed practices, farmers can reduce the environmental impact of feed production and use.

8. Education and Training:

Education and ongoing training are essential for the successful implementation of eco-friendly farming techniques. This involves:

- Farmer Education Programs: Providing farmers with access to information and training on sustainable farming practices. This can be through workshops, seminars, and online resources.
- Community Involvement: Engaging the local community in sustainability initiatives. This can include educational programs for schools and community groups to raise awareness about sustainable farming.
- Research and Development: Supporting research into new and improved sustainable farming techniques. Collaborating with academic institutions and agricultural organizations can lead to innovations that benefit both farmers and the environment.

By prioritizing education and training, farmers can stay informed about best practices and continuously improve their operations.

In conclusion, adopting eco-friendly farming techniques in crawfish farming is not only beneficial for the environment but also enhances the sustainability and profitability of farming operations. By integrating practices such as IPM, efficient water and waste management, habitat enhancement, energy efficiency, and sustainable feed practices, farmers can create a balanced and resilient farming system. Education and community involvement further support these efforts, ensuring that sustainable practices are maintained and continually improved. Through these collective efforts, crawfish farming can become a model of sustainable agriculture that supports both the environment and the community.

7.1.2 Waste Management

Effective waste management is a critical component of sustainable crawfish farming. As aquaculture systems, including crawfish farms, grow in scale and number, their environmental impacts, particularly in terms of waste production, become increasingly significant. This section will delve into various aspects of waste management specific to crawfish farming, including types of waste, their potential impacts, and strategies for efficient and environmentally friendly disposal or repurposing.

Types of Waste in Crawfish Farming

Crawfish farming generates several types of waste, each requiring distinct management approaches:

1. Organic Waste:

   - Uneaten Feed: Not all feed provided to crawfish is consumed. Uneaten feed decomposes and contributes to nutrient loads in the water.

   - Faecal Matter: Crawfish excrete waste, which accumulates in the water and sediment.

   - Dead Crawfish: Mortality is inevitable, and dead crawfish can become a significant source of organic waste.

2. Chemical Waste:

   - Fertilizers and Pesticides: These are sometimes used in pond preparation and maintenance, and their residues can accumulate in water and sediment.

   - Medication Residues: Treatments for diseases and parasites may leave chemical residues in the water.

3. Solid Waste:

- Debris and Trash: General farm operations can produce solid waste, including packaging materials, equipment parts, and general refuse.

Impacts of Waste on the Environment

The improper management of waste in crawfish farming can have several detrimental effects on the environment:

1. Eutrophication: Excess nutrients, primarily from uneaten feed and faecal matter, can lead to eutrophication. This process results in the overgrowth of algae, which depletes oxygen levels in the water, harming aquatic life.

2. Water Quality Degradation: High levels of organic waste can deteriorate water quality, making it unsuitable for crawfish and other aquatic organisms.

3. Sediment Accumulation: Waste products settle at the bottom of ponds, leading to sediment build-up. This can reduce the effective volume of the pond and affect water circulation.

4. Chemical Contamination: Residues from fertilizers, pesticides, and medications can contaminate water bodies, affecting non-target organisms and potentially entering the food chain.

Strategies for Effective Waste Management

To mitigate the environmental impacts of waste, crawfish farmers can adopt several strategies:

1. Feed Management:

   - Optimizing Feed Practices: Providing the right amount of feed at appropriate intervals reduces the amount of uneaten feed. Using high-quality feed that is easily digestible by crawfish can also minimize waste.

- Monitoring and Adjusting: Regular monitoring of feeding behavior and adjusting feeding rates accordingly helps prevent overfeeding.

2. Water Management:

- Water Exchange: Regular water exchange helps dilute waste concentrations in the pond. However, this must be managed carefully to avoid excessive water use and potential contamination of surrounding water bodies.

- Aeration: Using aerators can increase oxygen levels in the water, promoting the decomposition of organic matter and maintaining water quality.

3. Sediment Management:

- Periodic Sediment Removal: Removing accumulated sediment from the bottom of ponds periodically prevents excessive build-up and maintains pond capacity.

- Constructed Wetlands: Utilizing constructed wetlands adjacent to crawfish ponds can naturally filter and break down organic waste through plant and microbial activity.

4. Chemical Management:

- Responsible Use of Chemicals: Minimizing the use of fertilizers, pesticides, and medications, and opting for environmentally friendly alternatives where possible, reduces the risk of chemical contamination.

- Buffer Zones: Establishing buffer zones around ponds with vegetation can help filter runoff containing chemical residues before they reach natural water bodies.

5. Recycling and Repurposing Waste:

- Composting: Organic waste, such as dead crawfish and faecal matter, can be composted to produce nutrient-rich soil amendments.

- Biogas Production: Anaerobic digestion of organic waste can generate biogas, a renewable energy source, while producing digestate that can be used as fertilizer.

6. Monitoring and Record-Keeping:

  - Regular Testing: Conducting regular water quality tests helps in early detection of waste accumulation and allows for timely interventions.

  - Detailed Records: Keeping detailed records of feed inputs, chemical usage, and waste disposal practices helps in assessing the effectiveness of waste management strategies and making necessary adjustments.

Innovative Technologies and Approaches

Advancements in technology and innovative approaches are providing new tools and methods for waste management in crawfish farming:

1. Biofloc Technology:

  - Concept: Biofloc technology involves the cultivation of beneficial microbial communities within the aquaculture system. These microbes convert organic waste into microbial protein, which can be consumed by crawfish.

  - Benefits: This approach reduces waste accumulation, improves water quality, and provides an additional source of nutrition for crawfish.

2. Integrated Multi-Trophic Aquaculture (IMTA):

  - Concept: IMTA involves the co-cultivation of different species with complementary ecological functions. For example, combining crawfish farming with the cultivation of filter-feeding bivalves or aquatic plants that utilize waste nutrients.

  - Benefits: This system enhances nutrient recycling, reduces waste outputs, and increases overall productivity and sustainability of the farming operation.

3. Automated Feeding Systems:

- Concept: Automated feeding systems use sensors and software to optimize feeding schedules and amounts based on real-time data on crawfish behavior and pond conditions.

- Benefits: Precision feeding minimizes waste, enhances feed efficiency, and reduces labor costs.

4. Eco-Friendly Biofilters:

- Concept: Biofilters use natural or engineered materials to filter and break down waste products in the water. These can include gravel, sand, or specialized media designed to support beneficial microbial communities.

- Benefits: Biofilters help maintain water quality, reduce the need for water exchanges, and can be integrated into existing pond systems with relative ease.

Community and Policy Engagement

Effective waste management also involves engaging with the broader community and complying with environmental regulations:

1. Community Involvement:

- Education and Outreach: Educating local communities about sustainable practices and the importance of waste management can foster cooperation and support for environmentally friendly farming practices.

- Partnerships: Collaborating with local environmental organizations and research institutions can provide access to expertise, resources, and funding for waste management initiatives.

2. Regulatory Compliance:

- Adhering to Regulations: Compliance with local, regional, and national environmental regulations ensures that waste management practices meet legal standards and help protect natural resources.

- Participating in Certification Programs: Joining certification programs for sustainable aquaculture can provide recognition for environmentally responsible practices and access to premium markets.

Case Studies and Examples

Examining successful case studies can provide practical insights and inspiration for effective waste management:

1. Case Study 1: Louisiana Crawfish Farms:

   - Innovative Practices: Several crawfish farms in Louisiana have implemented biofloc systems and constructed wetlands to manage waste. These farms report improved water quality, reduced waste outputs, and enhanced crawfish growth rates.

   - Community Impact: Through community outreach programs, these farms have educated local residents about sustainable farming and have collaborated with schools to develop educational materials.

2. Case Study 2: Asian Integrated Farms:

   - IMTA Implementation: Integrated multi-trophic aquaculture systems combining crawfish with aquatic plants and bivalves have been successfully implemented in parts of Asia. These systems effectively recycle nutrients and reduce the environmental footprint of crawfish farming.

   - Economic Benefits: Farmers report increased revenue streams from the sale of additional crops, such as bivalves and aquatic plants, contributing to economic sustainability.

3. Case Study 3: European Sustainable Aquaculture Projects:

   - Policy and Innovation: European aquaculture projects often benefit from strong regulatory frameworks and innovative research. Several projects have demonstrated the efficacy of eco-friendly biofilters and automated feeding systems in reducing waste and improving overall farm efficiency.

- Collaborative Research: Collaboration between farmers, researchers, and policymakers has led to the development of best practices and guidelines for sustainable waste management in crawfish farming.

Conclusion

Effective waste management is essential for the sustainability and success of crawfish farming. By understanding the types and impacts of waste, and implementing a range of strategies—from optimizing feed practices and water management to adopting innovative technologies and engaging with the community—farmers can significantly reduce their environmental footprint. Sustainable waste management not only protects natural resources but also enhances the economic viability and social responsibility of crawfish farming operations. Through continuous improvement, innovation, and collaboration, the crawfish farming industry can achieve a balance between productivity and environmental stewardship, ensuring a sustainable future for both the industry and the ecosystems it relies on.

7.2 Economic Sustainability

Economic sustainability is a crucial component of successful crawfish farming. It ensures that the farming operation is not only profitable but also resilient in the face of economic challenges. This section focuses on strategies to achieve cost-efficiency, a key element in maintaining economic sustainability.

7.2.1 Cost-Efficiency Strategies

Cost-efficiency in crawfish farming involves optimizing the use of resources, reducing unnecessary expenditures, and maximizing productivity. Implementing cost-efficient strategies can significantly enhance the profitability and sustainability of your farming operation. Below are several effective strategies for achieving cost-efficiency in crawfish farming:

1. Optimizing Feed Usage

Feed is one of the largest recurring costs in crawfish farming. Effective feed management can lead to substantial cost savings:

- Accurate Feeding: Ensure that the crawfish are fed the right amount of food. Overfeeding can lead to wastage and increased costs, while underfeeding can stunt growth and reduce yield. Use feeding charts and observe the behavior of crawfish to determine the optimal feeding amounts.

- Quality Over Quantity: Invest in high-quality feed that is rich in nutrients. Although high-quality feed may be more expensive upfront, it promotes better growth and health of the crawfish, leading to higher yields and lower mortality rates.

- Alternative Feed Sources: Explore alternative feed sources, such as agricultural by-products or formulated feeds that are cost-effective and nutritionally adequate.

2. Efficient Water Management

Water management is critical for maintaining the health of crawfish and minimizing costs associated with water usage and treatment:

- Water Recycling: Implement systems to recycle water within the farm. Recycled water can reduce the need for fresh water and lower costs related to water procurement and waste treatment.
- Pond Design: Design ponds to maximize water retention and minimize evaporation. Properly designed ponds can reduce the frequency of water replacement and associated costs.
- Monitoring Water Quality: Regularly monitor water quality parameters such as pH, temperature, and oxygen levels. Maintaining optimal water quality reduces the risk of disease outbreaks and enhances the overall productivity of the farm.

3. Energy Efficiency

Energy costs can be a significant portion of operational expenses. Implementing energy-efficient practices can lead to substantial savings:

- Renewable Energy Sources: Invest in renewable energy sources such as solar or wind power. Although the initial investment may be high, renewable energy can significantly reduce long-term energy costs.

- Energy-Efficient Equipment: Use energy-efficient pumps, aerators, and other equipment. Regular maintenance and timely replacement of old equipment can prevent energy wastage.

- Automation: Automate processes such as feeding, water circulation, and monitoring. Automation reduces labor costs and ensures consistent and efficient operation.

4. Labor Management

Efficient labor management is essential for controlling operational costs:

- Training: Invest in the training of employees to enhance their skills and efficiency. Well-trained staff can perform tasks more effectively, reducing errors and increasing productivity.

- Optimal Staffing Levels: Assess the labor requirements for different stages of the farming cycle and adjust staffing levels accordingly. Avoid overstaffing during periods of low activity and ensure sufficient labor during peak periods.

- Incentive Programs: Implement incentive programs to motivate employees. Performance-based incentives can boost productivity and reduce labor turnover, which in turn lowers recruitment and training costs.

5. Infrastructure Investment

Strategic investment in infrastructure can enhance efficiency and reduce long-term costs:

- Durable Materials: Use durable and low-maintenance materials for constructing ponds, buildings, and other structures. This reduces the frequency of repairs and replacements.

- Technological Upgrades: Invest in modern technology for monitoring and managing the farm. Technologies such as remote sensors, automated feeders, and data analytics can improve efficiency and reduce operational costs.

- Scalability: Design infrastructure with scalability in mind. Scalable infrastructure allows for gradual expansion without significant additional investment, providing flexibility to adapt to market demands.

6. Disease Prevention and Health Management

Preventing diseases and maintaining the health of crawfish is essential for cost-efficiency:

- Biosecurity Measures: Implement stringent biosecurity measures to prevent the introduction and spread of diseases. This includes controlling access to the farm, using disinfectants, and monitoring the health of crawfish regularly.

- Vaccination and Medication: Use vaccinations and medications judiciously to prevent and treat diseases. While this may incur some costs, it is often cheaper than dealing with a full-blown disease outbreak.

- Healthy Stock: Source healthy and disease-free stock from reputable suppliers. Healthy stock is less likely to require medical interventions and tends to have higher survival and growth rates.

7. Market Strategies

Effective market strategies can help maximize revenue and reduce financial risks:

- Diversification: Diversify the products offered by the farm. In addition to selling live crawfish, consider value-added products such as processed crawfish meat, crawfish bait, or crawfish shells for fertilizer.

- Market Research: Conduct regular market research to understand trends and demand. Adjust production and marketing strategies based on market insights to capture higher prices and expand market reach.

- Direct Sales: Explore direct sales channels such as farmers' markets, online sales, and local restaurants. Direct sales can often yield higher prices compared to selling through intermediaries.

8. Financial Planning and Risk Management

Comprehensive financial planning and risk management are crucial for economic sustainability:

- Budgeting: Develop detailed budgets that account for all aspects of the farming operation. Regularly review and adjust budgets to ensure alignment with financial goals.

- Insurance: Obtain insurance to protect against risks such as natural disasters, disease outbreaks, and market fluctuations. Insurance can provide financial stability and mitigate losses.

- Diversified Income Streams: Explore diversified income streams, such as agritourism, consulting services, or educational workshops. Diversified income can buffer against market volatility and enhance overall financial stability.

9. Sustainable Practices

Adopting sustainable practices not only benefits the environment but can also lead to cost savings:

- Integrated Pest Management: Use integrated pest management (IPM) techniques to control pests. IPM reduces reliance on chemical pesticides, which can be expensive and harmful to the environment.

- Organic Farming: Consider organic farming practices that reduce the need for synthetic inputs. Organic products can often command higher prices in the market, enhancing profitability.

- Waste Utilization: Implement systems to utilize waste products, such as composting crawfish shells or using pond sediment as fertilizer. Waste utilization can reduce disposal costs and generate additional revenue.

In conclusion, achieving cost-efficiency in crawfish farming requires a holistic approach that encompasses various aspects of farm management. By optimizing feed usage, managing water and energy efficiently, investing in infrastructure, preventing diseases, implementing effective market strategies, and planning financially, farmers can enhance the economic sustainability of their operations. Embracing sustainable practices not only contributes to cost-efficiency but also promotes a healthier environment and long-term viability of the crawfish farming industry.

7.2.2 Long-Term Planning

Long-term planning is crucial for the economic sustainability of any crawfish farming operation. It involves forecasting future trends, setting strategic goals, and implementing measures to ensure the farm's viability and profitability over extended periods. In this section, we delve into the key considerations and strategies for effective long-term planning in crawfish farming.

Understanding Market Trends

One of the fundamental aspects of long-term planning in crawfish farming is staying informed about market trends. Markets for crawfish can fluctuate due to various factors such as consumer preferences, economic conditions, and global supply. Farmers need to conduct thorough market research and analysis to anticipate these trends accurately.

Monitoring market demand and prices allows farmers to make informed decisions about production levels, timing of harvests, and pricing strategies. This proactive approach minimizes risks associated with market volatility and helps in maximizing profitability over the long term.

Investment in Infrastructure and Technology

Investing in infrastructure and technology is another critical component of long-term planning. As crawfish farming evolves, so do the techniques and tools available to improve efficiency and reduce costs. Upgrading ponds, installing efficient water management systems, and adopting advanced feeding and harvesting technologies can significantly enhance productivity and sustainability.

For example, automated feeding systems can ensure consistent nutrition for crawfish, while modern pond liners and aerators improve water quality and minimize environmental impact. These investments not only boost operational efficiency but also contribute to the overall economic viability of the farm in the long run.

Diversification and Risk Management

Long-term planning should also encompass diversification and risk management strategies. Diversifying the product offerings beyond live crawfish, such as processed crawfish products or value-added items, can create additional revenue streams and reduce dependency on fluctuating market prices.

Moreover, effective risk management involves identifying potential threats to the farm's profitability and implementing measures to mitigate them. This includes hedging strategies for input costs, insurance coverage for unforeseen events like natural disasters, and contingency plans for disease outbreaks or regulatory changes.

Sustainable Growth and Expansion

Planning for sustainable growth and expansion is essential for the long-term success of a crawfish farm. This involves evaluating opportunities for scaling up operations while maintaining environmental sustainability and financial stability.

Expansion may include acquiring additional acreage for pond development, increasing production capacity, or exploring new markets both domestically and internationally. However, expansion plans must be carefully balanced with considerations for resource availability, environmental impact assessments, and financial feasibility studies.

Succession Planning

Finally, succession planning is a critical aspect of long-term planning for family-owned or generational crawfish farms. It involves outlining a clear roadmap for transferring ownership and management responsibilities to the next generation or new owners seamlessly. Succession planning ensures continuity in operations, preserves accumulated knowledge and expertise, and safeguards the farm's legacy for future generations.

In conclusion, effective long-term planning in crawfish farming integrates market intelligence, strategic investments, risk management, sustainable growth practices, and succession planning. By adopting a proactive approach to these considerations, crawfish farmers can navigate challenges, capitalize on opportunities, and build a resilient and economically sustainable enterprise over time.

7.3 Social Responsibility

7.3.1 Community Involvement

Social responsibility in crawfish farming extends beyond environmental and economic considerations to encompass active engagement with the local community. This involvement is crucial not only for fostering goodwill but also for ensuring the sustainability and resilience of the farm's operations. Community involvement can take various forms, each contributing to the farm's integration into the local social fabric and its role as a responsible corporate entity.

Building Strong Community Relationships

Establishing and maintaining strong relationships with the local community is fundamental to the success of any crawfish farm. These relationships are built on trust, transparency, and mutual benefit. Farm operators should strive to:

- Communicate Effectively: Regular and clear communication with community members helps in addressing concerns, sharing updates on farm activities, and soliciting feedback.

- Engage Proactively: Actively seek opportunities to engage with community members through local events, workshops, or informational sessions about crawfish farming practices.

- Support Local Initiatives: Partner with local schools, nonprofits, or community organizations on projects related to education, environmental conservation, or economic development.

Environmental Education and Awareness

Educating the community about the environmental impact and benefits of crawfish farming can foster understanding and support. This can include:

- Educational Tours: Offer guided tours of the farm to local schools, groups, or interested individuals to showcase sustainable farming practices and the ecological importance of crawfish habitats.

- Workshops and Seminars: Organize workshops on topics such as water quality management, habitat conservation, and the role of crawfish farming in local ecosystems.

- Demonstrate Best Practices: Implement visible environmental stewardship practices that the community can observe, such as wetland restoration, biodiversity enhancement, or water conservation efforts.

Promoting Local Economic Development

Crawfish farming can significantly contribute to local economies through job creation and supporting ancillary industries. To maximize these benefits and foster economic sustainability:

- Local Employment: Prioritize hiring locally whenever possible, offering training and career development opportunities to community members.

- Supply Chain Support: Source inputs and supplies from local businesses to stimulate economic growth and strengthen interdependencies within the community.

- Market Integration: Partner with local markets, restaurants, and retailers to promote locally grown crawfish, thereby boosting the agricultural sector and increasing consumer awareness of regional products.

Addressing Community Concerns

Acknowledge and address any concerns or challenges voiced by community members promptly and transparently. Key areas of concern may include:

- Noise and Odor Management: Implement measures to mitigate noise and odor impacts on nearby residents through technological advancements or strategic farm layout.

- Water Use and Quality: Monitor and manage water usage responsibly, ensuring minimal impact on local water resources and maintaining water quality standards.

- Traffic and Infrastructure: Collaborate with local authorities to manage traffic flow and infrastructure improvements around the farm, especially during peak seasons.

Conclusion

Community involvement is not just a corporate responsibility but a strategic imperative for crawfish farms aiming for long-term sustainability and positive societal impact. By engaging actively with the community, promoting environmental education, supporting local economic development, and addressing concerns transparently, crawfish farms can foster mutual respect and cooperation while securing their social license to operate.

7.3.2 Ethical Farming Practices

Ethical farming practices in crawfish farming are not just about meeting regulatory requirements but also about ensuring humane treatment of the animals, promoting worker well-being, and fostering sustainable relationships within the community.

Animal Welfare

Ethical considerations in crawfish farming extend to the treatment and welfare of the crustaceans themselves. Crawfish are sensitive creatures that require careful handling to minimize stress and ensure their well-being throughout their lifecycle. Ethical practices in crawfish farming include:

- Minimizing Stress: Implementing handling techniques that reduce stress during harvesting and transport.
- Optimal Conditions: Maintaining water quality and appropriate stocking densities to support crawfish health and growth.
- Avoiding Cruel Practices: Avoiding practices such as overcrowding or neglect that could harm the crawfish.

In ethical crawfish farming, farmers prioritize the health and comfort of their stock, recognizing that humane treatment not only aligns with ethical standards but also contributes to better product quality.

Worker Well-Being

Social responsibility in crawfish farming also encompasses the well-being of those involved in the process, including farm workers and laborers. Ethical considerations here involve:

- Safe Working Conditions: Providing a safe working environment that adheres to health and safety regulations.

- Fair Compensation: Ensuring fair wages and appropriate compensation for all workers involved in the farming operations.

- Training and Development: Offering training opportunities to improve skills and knowledge related to crawfish farming practices.

By prioritizing worker well-being, ethical crawfish farmers contribute positively to their local communities and demonstrate a commitment to responsible business practices.

Community Engagement

Ethical crawfish farming goes beyond the farm gates and extends into the community. Farmers engage with local communities in various ways, including:

- Environmental Stewardship: Participating in local conservation efforts and promoting environmental awareness among community members.

- Education Programs: Hosting farm visits or educational sessions to educate the public about crawfish farming practices and sustainability.

- Supporting Local Initiatives: Contributing to community development projects or local charities to foster positive relationships and support social causes.

Community involvement not only enhances the reputation of crawfish farms but also strengthens community ties and promotes mutual understanding and support.

Transparency and Accountability

Ethical crawfish farmers prioritize transparency in their operations, ensuring accountability to stakeholders such as consumers, regulatory bodies, and local communities. Key aspects of transparency include:

- Traceability: Providing clear information about the origin and production practices of crawfish products.

- Compliance: Adhering to legal requirements and industry standards related to environmental sustainability and animal welfare.

- Feedback Mechanisms: Welcoming feedback from stakeholders and using it to improve farming practices continuously.

By maintaining transparency and accountability, ethical crawfish farmers build trust with consumers and stakeholders, demonstrating their commitment to ethical standards and sustainable practices.

Conclusion

Ethical farming practices in crawfish farming encompass a range of considerations, from animal welfare and worker well-being to community engagement and transparency. By embracing ethical principles, crawfish farmers not only contribute to sustainable agriculture but also promote social responsibility and enhance the overall quality and integrity of their products. As crawfish farming continues to evolve, ethical practices will play an increasingly important role in shaping the industry's future.

PART VIII
Advanced Crawfish Farming Techniques

8.1 Innovative Breeding Methods

8.1.1 Genetic Selection

Introduction

Genetic selection is a cornerstone in the field of advanced aquaculture, providing a scientific approach to improving the quality and productivity of crawfish farming. By selecting and breeding individuals with desirable traits, farmers can enhance the growth rate, disease resistance, and overall yield of their crawfish populations. This section delves into the principles, techniques, and benefits of genetic selection in crawfish farming, offering a comprehensive guide for beginners and experienced farmers alike.

Principles of Genetic Selection

Genetic selection involves identifying and breeding individuals with specific genetic traits that are deemed advantageous for farming. These traits can include rapid growth, resistance to diseases, better feed conversion ratios, and improved reproductive capabilities. The process relies on understanding the genetic makeup of crawfish and using this knowledge to inform breeding decisions.

Heritability

One fundamental concept in genetic selection is heritability, which measures how much of the variation in a trait is due to genetic factors. High heritability indicates that selective breeding will be more effective, as the desired traits are likely to be passed on to the next generation. For crawfish, traits such as growth rate and disease resistance often exhibit moderate to high heritability.

Genetic Diversity

Maintaining genetic diversity within the breeding population is crucial to avoid inbreeding depression, which can lead to reduced fitness and increased susceptibility to diseases. Genetic diversity ensures a broad genetic base, allowing for the selection of robust individuals and the introduction of new beneficial traits over time.

Techniques in Genetic Selection

Several techniques are employed in genetic selection to enhance crawfish populations effectively. These methods range from traditional selective breeding to modern molecular techniques.

Traditional Selective Breeding

Traditional selective breeding involves choosing the best-performing individuals based on phenotypic traits and using them as broodstock. Farmers typically select crawfish that demonstrate superior growth rates, larger size, and greater resilience to environmental stressors. This method, while straightforward, relies heavily on observable traits, which may not always accurately reflect the genetic potential of the individuals.

1. Mass Selection: This technique involves selecting a large number of individuals based on desirable traits and breeding them. Over successive generations, the average performance of the population improves. Mass selection is simple and cost-effective but may be less precise compared to other methods.

2. Family Selection: In family selection, the performance of entire families (groups of related individuals) is evaluated. The best families are then chosen for breeding. This

method provides more control over the genetic makeup of the population and helps manage inbreeding.

3. Individual Selection: Individual selection focuses on selecting the best-performing individuals within a family or population. This method is highly precise but can be labor-intensive and requires detailed record-keeping.

Marker-Assisted Selection (MAS)

Marker-assisted selection leverages molecular markers—specific DNA sequences associated with desirable traits—to guide breeding decisions. By identifying and selecting individuals with the desired genetic markers, farmers can accelerate the breeding process and achieve more consistent results.

1. Identification of Markers: The first step in MAS is identifying DNA markers linked to important traits. This is achieved through genetic mapping and association studies, which correlate specific genetic sequences with phenotypic traits.

2. Screening and Selection: Once markers are identified, crawfish can be screened for these markers using molecular techniques such as polymerase chain reaction (PCR). Individuals with the desired markers are selected for breeding, ensuring that the beneficial traits are passed on to the next generation.

3. Advantages: MAS allows for early selection of individuals, even before traits are visibly expressed. This reduces the breeding cycle time and improves the accuracy of selection. It also enables the selection of traits that are difficult to measure phenotypically, such as disease resistance.

Genomic Selection

Genomic selection represents the cutting edge of genetic improvement in aquaculture. This technique uses genome-wide genetic information to predict the breeding value of individuals. By analyzing thousands of genetic markers spread across the genome, genomic selection provides a comprehensive understanding of an individual's genetic potential.

1. Genomic Predictions: The process begins with the creation of a reference population, where both phenotypic and genotypic data are collected. Statistical models are then developed to predict the breeding value of individuals based on their genomic data.

2. Implementation: Farmers can use genomic predictions to select the best individuals for breeding. This method is highly efficient and can significantly enhance the rate of genetic gain compared to traditional methods.

3. Benefits: Genomic selection offers several advantages, including higher accuracy, the ability to select for complex traits, and reduced generation intervals. It is particularly useful for traits with low heritability or those that are costly or time-consuming to measure.

Benefits of Genetic Selection

Implementing genetic selection in crawfish farming offers numerous benefits that can substantially improve farm productivity and profitability.

Enhanced Growth Rates

Selective breeding for rapid growth can lead to faster production cycles and higher yields. By choosing individuals that grow quickly and efficiently, farmers can shorten the time to market size, increasing overall production capacity.

Improved Disease Resistance

Disease outbreaks are a significant challenge in aquaculture. Genetic selection for disease-resistant traits can reduce mortality rates and the need for chemical treatments, leading to healthier populations and lower operational costs.

Better Feed Conversion

Efficient feed conversion is crucial for the economic viability of crawfish farming. By selecting individuals that convert feed into body mass more effectively, farmers can reduce feed costs and enhance the sustainability of their operations.

Increased Reproductive Performance

Enhanced reproductive traits, such as higher fecundity and improved hatch rates, contribute to more robust and resilient crawfish populations. This ensures a stable supply of juveniles for stocking ponds and supports the long-term sustainability of the farming operation.

Challenges and Considerations

While genetic selection offers significant advantages, it also presents several challenges that farmers must navigate to achieve success.

Managing Inbreeding

Inbreeding can lead to a decrease in genetic diversity and an increase in the expression of deleterious traits. To manage inbreeding, farmers should maintain a large and genetically diverse breeding population, use rotational breeding strategies, and periodically introduce new genetic material from external sources.

Balancing Traits

Focusing on a single trait, such as growth rate, can sometimes lead to unintended consequences, such as reduced disease resistance or poorer reproductive performance. A balanced approach to selection, considering multiple traits simultaneously, is essential for overall population health and productivity.

Technological and Financial Barriers

Advanced techniques like MAS and genomic selection require significant technological infrastructure and expertise. The initial costs of implementing these technologies can be high, posing a barrier for small-scale farmers. However, collaborative efforts and access to shared resources, such as genetic testing facilities, can help mitigate these challenges.

Future Directions

The field of genetic selection in crawfish farming is continually evolving, with ongoing research aimed at uncovering new genetic markers, improving selection accuracy, and developing more efficient breeding strategies. As technology advances, the integration of artificial intelligence and machine learning in genetic selection holds promise for further enhancing the precision and efficiency of breeding programs.

Conclusion

Genetic selection is a powerful tool in crawfish farming, offering the potential to significantly improve growth rates, disease resistance, feed conversion, and reproductive performance. By understanding and implementing innovative breeding methods, farmers can achieve higher yields and more sustainable operations. While challenges remain, ongoing research and technological advancements continue to pave the way for a more productive and resilient crawfish farming industry.

8.1.2 Hybrid Breeding

Hybrid breeding is a sophisticated and highly beneficial technique in crawfish farming, aimed at enhancing desirable traits such as growth rate, disease resistance, and environmental adaptability. This method involves crossbreeding two or more genetically distinct populations to produce offspring, or hybrids, that exhibit superior qualities compared to their parent strains. Hybrid breeding in crawfish is an advanced approach that requires careful planning, execution, and ongoing management to realize its full potential.

The Concept of Hybrid Breeding

Hybrid breeding leverages the principle of heterosis, or hybrid vigor, which posits that hybrid offspring often perform better than their parents. This phenomenon is due to the combination of diverse genetic materials that can result in enhanced vitality, increased growth rates, and improved resilience to environmental stresses and diseases.

In the context of crawfish farming, hybrid breeding aims to combine the best traits of different crawfish species or strains. For example, a strain known for its rapid growth rate might be crossbred with another that has a high resistance to common diseases. The resulting hybrids could potentially exhibit both rapid growth and disease resistance, making them more profitable and easier to manage for farmers.

Steps in Hybrid Breeding

1. Selection of Parent Strains: The first step in hybrid breeding involves selecting parent strains with desirable traits. This requires a thorough understanding of the genetic makeup and performance characteristics of different crawfish strains. Farmers and researchers must evaluate factors such as growth rates, feed efficiency, disease resistance, and adaptability to local environmental conditions.

2. Controlled Breeding: Once the parent strains are selected, controlled breeding techniques are employed to produce hybrids. This typically involves artificial insemination or managed mating in controlled environments to ensure that the desired genetic traits are passed on to the offspring. Detailed records of breeding pairs and their traits are crucial for monitoring and evaluating the success of the hybridization process.

3. Raising the Hybrids: The hybrid offspring are raised under optimal conditions to ensure their health and growth. This stage involves regular monitoring of water quality, feeding practices, and health assessments to maximize the potential of the hybrids. Farmers need to pay close attention to the specific needs of the hybrid crawfish, as they might differ slightly from their parent strains.

4. Evaluation and Selection: Not all hybrids will exhibit the desired traits, so a rigorous evaluation process is necessary. Farmers must assess the performance of hybrid crawfish in terms of growth rate, survival rate, feed conversion efficiency, and disease resistance. The best-performing individuals are selected for further breeding or for commercial production.

5. Continuous Improvement: Hybrid breeding is an ongoing process. Continuous evaluation and selection help in refining the genetic pool and improving the overall performance of hybrid crawfish. This iterative process can lead to significant advancements in the quality and productivity of crawfish farms over time.

Advantages of Hybrid Breeding

1. Enhanced Growth Rates: One of the most significant advantages of hybrid breeding is the potential for enhanced growth rates. Hybrids often grow faster than their parent strains, reducing the time required to reach market size and increasing the overall productivity of the farm.

2. Improved Disease Resistance: By combining genetic traits from different strains, hybrids can exhibit greater resistance to common diseases. This reduces mortality rates and the need for chemical treatments, leading to healthier crawfish and a more sustainable farming practice.

3. Greater Environmental Adaptability: Hybrids can be more adaptable to varying environmental conditions, making them more resilient to changes in water quality,

temperature, and other factors. This adaptability can result in more stable production levels and reduce the impact of environmental stressors.

4. Better Feed Conversion Efficiency: Hybrids often exhibit better feed conversion ratios, meaning they convert feed into body mass more efficiently. This can lower feeding costs and improve the overall profitability of the farm.

Challenges and Considerations

While hybrid breeding offers numerous benefits, it also presents several challenges that farmers must address:

1. Genetic Management: Managing the genetic diversity of hybrid populations requires careful planning and record-keeping. Inbreeding can lead to reduced vigor and performance, so maintaining a broad genetic base is essential.

2. Initial Investment: The initial costs of establishing a hybrid breeding program can be high, including the acquisition of parent strains, setting up controlled breeding environments, and ongoing monitoring and evaluation.

3. Market Acceptance: Some markets may have preferences for specific crawfish strains, and introducing hybrids might require efforts to educate consumers and market the benefits of hybrid crawfish.

4. Regulatory Considerations: There may be regulatory requirements or restrictions related to the introduction of hybrid species, particularly if they are to be farmed in open water systems where they might interact with wild populations.

Case Studies and Success Stories

Several successful hybrid breeding programs in aquaculture can provide insights and inspiration for crawfish farmers:

1. Tilapia: Hybrid tilapia, such as the widely farmed Nile and Mozambique tilapia hybrids, have shown remarkable improvements in growth rates, disease resistance, and adaptability. The success of these programs highlights the potential benefits of hybrid breeding in aquaculture.

2. Shrimp: In shrimp farming, hybrids between different species or strains have led to improved growth rates, better survival rates, and enhanced resistance to diseases like white spot syndrome and early mortality syndrome.

3. Salmon: Hybrid salmon have been developed to exhibit better growth rates and disease resistance, contributing to more sustainable and profitable salmon farming operations.

Future Prospects

The future of hybrid breeding in crawfish farming holds great promise. Advances in genetic research, including genome sequencing and gene editing technologies, could further enhance the effectiveness of hybrid breeding programs. These technologies can help identify specific genes associated with desirable traits, enabling more precise and targeted breeding strategies.

Moreover, as the global demand for seafood continues to rise, the development of high-performing hybrid crawfish can play a crucial role in meeting this demand sustainably. Hybrid crawfish farming can contribute to food security, economic development, and the preservation of natural crawfish populations by reducing the pressure on wild stocks.

Conclusion

Hybrid breeding is a powerful tool in the advancement of crawfish farming, offering numerous benefits in terms of growth rates, disease resistance, environmental adaptability, and feed efficiency. While there are challenges to be addressed, the potential rewards make it a worthwhile endeavor for forward-thinking farmers. By embracing innovative breeding methods and leveraging advancements in genetic research, the crawfish farming industry can achieve new levels of productivity and sustainability.

8.2 Technology in Crawfish Farming

Advancements in technology have revolutionized various sectors, and aquaculture is no exception. In crawfish farming, the integration of technology has significantly enhanced productivity, efficiency, and sustainability. This section will explore the key technological advancements in crawfish farming, focusing particularly on automated feeding systems.

8.2.1 Automated Feeding Systems

Automated feeding systems represent a significant leap forward in the management of crawfish farms. These systems are designed to deliver precise amounts of feed to crawfish at scheduled intervals, reducing waste and ensuring optimal growth conditions. This section delves into the different aspects of automated feeding systems, their benefits, and the considerations for implementing them in a crawfish farming operation.

1. Introduction to Automated Feeding Systems

Automated feeding systems in aquaculture have been developed to streamline the feeding process, reduce labor costs, and improve the overall efficiency of feed usage. These systems range from simple timer-based feeders to sophisticated devices that can be controlled remotely and adjusted based on real-time data.

2. Components of Automated Feeding Systems

Automated feeding systems typically consist of several key components:

- Feed Storage and Dispensing Unit: This is where the feed is stored and from where it is dispensed. These units come in various sizes and capacities, depending on the scale of the farming operation.

- Control Unit: This is the brain of the automated feeding system. It controls when and how much feed is dispensed based on pre-programmed settings or real-time inputs.

- Dispersion Mechanism: This component ensures that the feed is evenly distributed throughout the farming area. It can be a series of tubes, conveyors, or rotors that spread the feed across the water surface.

- Sensors and Monitors: Modern systems include sensors that monitor various parameters such as water quality, crawfish activity, and feed consumption. This data helps in fine-tuning the feeding process.

3. Types of Automated Feeding Systems

There are several types of automated feeding systems available, each with its own set of features and benefits:

- Timer-Based Feeders: These are the simplest form of automated feeders, where feed is dispensed at preset times. They are easy to set up and use but lack the sophistication of more advanced systems.

- Demand Feeders: These feeders dispense feed based on the demand of the crawfish. They have sensors that detect when the crawfish are actively feeding and adjust the feed delivery accordingly.

- Intelligent Feeders: These are advanced systems that use data from various sensors to adjust feeding schedules and quantities in real-time. They can be integrated with other farm management systems and controlled remotely.

4. Benefits of Automated Feeding Systems

Implementing automated feeding systems in crawfish farming offers numerous advantages:

- Increased Efficiency: Automated systems can feed crawfish more consistently and accurately than manual feeding, ensuring optimal growth and reducing feed wastage.

- Labor Savings: These systems reduce the need for manual labor, allowing farm operators to focus on other important tasks. This is particularly beneficial for large-scale operations.

- Improved Feed Management: Automated systems can adjust feed quantities based on real-time data, preventing overfeeding or underfeeding. This not only saves costs but also promotes healthier crawfish.

- Enhanced Monitoring: The integration of sensors and monitors provides valuable data on crawfish behavior and environmental conditions, helping farmers make informed decisions.

- Scalability: Automated feeding systems can be easily scaled up or down, making them suitable for farms of all sizes.

5. Challenges and Considerations

While automated feeding systems offer many benefits, there are also challenges and considerations to keep in mind:

- Initial Investment: The cost of purchasing and installing an automated feeding system can be significant. However, this is often offset by the long-term savings in labor and feed costs.

- Maintenance: Like any technological system, automated feeders require regular maintenance to ensure they function correctly. This includes cleaning, calibration, and occasional repairs.

- Integration with Existing Systems: For farms that already have established manual feeding routines, transitioning to an automated system can require significant adjustments and training.

- Technical Expertise: Operating and maintaining these systems may require technical knowledge and skills that the existing farm staff might need to acquire.

6. Case Studies and Examples

Several crawfish farms have successfully implemented automated feeding systems, demonstrating their efficacy and benefits. For example:

- Case Study 1: A large-scale crawfish farm in Louisiana implemented an intelligent feeding system integrated with water quality sensors. The system reduced feed wastage by 20% and improved the overall growth rate of the crawfish, resulting in higher yields and profits.

- Case Study 2: A small family-owned farm in Texas adopted a timer-based feeding system. The automation allowed the farmers to reduce labor costs significantly and allocate more time to other aspects of farm management.

7. Future Trends in Automated Feeding

The future of automated feeding in crawfish farming looks promising, with several trends emerging:

- Integration with IoT: The Internet of Things (IoT) is set to play a significant role in the future of automated feeding systems. IoT-enabled feeders can communicate with other farm equipment and provide real-time data to a centralized management system.

- AI and Machine Learning: Artificial intelligence (AI) and machine learning algorithms can analyze data from sensors and predict feeding patterns, further optimizing feed usage and improving farm productivity.

- Sustainability: Future automated feeding systems will likely focus on sustainability, using eco-friendly materials and energy-efficient designs to minimize the environmental impact of crawfish farming.

8. Conclusion

Automated feeding systems are a valuable addition to modern crawfish farming, offering numerous benefits including increased efficiency, labor savings, and improved feed management. While there are challenges and considerations to address, the potential for enhanced productivity and sustainability makes them an attractive investment for crawfish farmers. As technology continues to advance, these systems will become even more sophisticated, further transforming the landscape of aquaculture.

8.2.2 Water Quality Monitoring Technologies

Water quality is a critical factor in the success of any aquaculture operation, including crawfish farming. Poor water quality can lead to decreased growth rates, increased susceptibility to diseases, and higher mortality rates among crawfish. Therefore, implementing advanced water quality monitoring technologies is essential to maintain optimal conditions and ensure the health and productivity of the crawfish. This section explores the various water quality parameters that need monitoring, the technologies available for monitoring these parameters, and the benefits of integrating these technologies into crawfish farming operations.

Importance of Water Quality Monitoring

Crawfish are highly sensitive to changes in their aquatic environment. Key water quality parameters that affect crawfish health include temperature, dissolved oxygen, pH, ammonia, nitrite, nitrate, and hardness. Monitoring these parameters helps farmers to detect and address issues before they become detrimental to the crawfish population. Maintaining optimal water quality ensures that the crawfish can thrive, grow efficiently, and be free from diseases.

Key Water Quality Parameters

1. Temperature: Crawfish are ectothermic, meaning their body temperature is regulated by the surrounding water. Optimal temperature ranges are crucial for their metabolic processes, growth, and reproduction. Typically, the ideal temperature range for crawfish farming is between 20-25°C (68-77°F). Temperatures outside this range can cause stress and adversely affect their growth and survival.

2. Dissolved Oxygen (DO): Oxygen is vital for the respiration of crawfish. Low DO levels can lead to hypoxia, causing stress, reduced growth, and even death. It is recommended to maintain DO levels above 3 mg/L to ensure the well-being of the crawfish.

3. pH: The pH level of the water affects the biological processes of crawfish. A pH range of 6.5-8.5 is considered optimal. Levels outside this range can cause physiological stress and impact the survival of the crawfish.

4. Ammonia, Nitrite, and Nitrate: These nitrogenous compounds are toxic to crawfish at high concentrations. Ammonia (NH_3) is particularly harmful as it can affect gill function and lead to respiratory distress. Regular monitoring and management are necessary to keep these compounds at safe levels.

5. Hardness: Water hardness, determined by the concentration of calcium and magnesium, influences shell formation and molting in crawfish. Adequate levels of these minerals are essential for the development of a strong exoskeleton.

Technologies for Monitoring Water Quality

Several advanced technologies are available for monitoring water quality in crawfish farming. These technologies range from simple handheld devices to sophisticated automated systems that provide real-time data and alerts.

1. Handheld Digital Meters:

 - Portable Multi-Parameter Meters: These devices can measure multiple water quality parameters such as temperature, DO, pH, and conductivity. They are convenient for spot-checking and quick assessments.

- Specific Parameter Meters: Handheld meters designed for individual parameters (e.g., DO meters, pH meters) provide accurate and reliable measurements. They are often used for regular monitoring and calibration of automated systems.

2. Automated Water Quality Monitoring Systems:

  - Integrated Sensor Systems: These systems consist of various sensors placed in the water body to continuously measure parameters like temperature, DO, pH, and ammonia. Data is collected and transmitted to a central unit for analysis.

  - Data Loggers: Data loggers record water quality parameters over time, allowing farmers to track changes and trends. They are useful for identifying long-term patterns and potential issues.

  - Real-Time Monitoring and Alert Systems: Advanced systems provide real-time data and can send alerts via SMS or email if parameters exceed safe limits. This allows for immediate intervention to prevent adverse effects on the crawfish.

3. Remote Sensing Technologies:

  - Satellite and Aerial Imaging: While more commonly used in large-scale aquaculture operations, these technologies can provide valuable information on water quality and environmental conditions. They can help in assessing parameters like temperature and chlorophyll concentrations, indicating algal blooms.

4. Smart Aquaculture Systems:

  - IoT-Based Monitoring: Internet of Things (IoT) devices can connect various sensors to a central network, providing comprehensive monitoring and control. These systems enable farmers to manage water quality remotely and make data-driven decisions.

  - Machine Learning and AI: Advanced algorithms can analyze water quality data to predict trends and suggest optimal management practices. AI can help in optimizing feeding schedules, aeration, and other critical operations.

Benefits of Advanced Water Quality Monitoring Technologies

1. Improved Efficiency: Automated systems reduce the need for manual monitoring, saving time and labor. Real-time data allows for prompt adjustments to maintain optimal water conditions.

2. Enhanced Precision: Continuous monitoring provides more accurate and consistent data compared to periodic manual checks. This precision helps in fine-tuning management practices.

3. Early Detection of Issues: Real-time alerts enable farmers to identify and address water quality problems before they escalate, preventing losses and ensuring the health of the crawfish.

4. Data-Driven Decisions: Historical data and trend analysis support informed decision-making, leading to better resource management and improved farm performance.

5. Sustainability: Efficient water quality management contributes to the sustainability of crawfish farming by minimizing environmental impacts and promoting the health of the ecosystem.

Implementation and Integration

Integrating advanced water quality monitoring technologies into crawfish farming requires careful planning and investment. Here are some steps to consider:

1. Assessment of Needs: Determine the specific water quality parameters that need monitoring based on the characteristics of the farming environment and the species of crawfish being cultivated.

2. Selection of Technology: Choose the appropriate monitoring tools and systems that meet the needs and budget of the farm. Consider factors such as ease of use, accuracy, durability, and cost.

3. Installation and Calibration: Proper installation and calibration of sensors and monitoring devices are crucial for obtaining accurate data. Regular maintenance and calibration ensure the reliability of the system.

4. Training and Education: Farmers and farm workers should be trained on how to use and maintain the monitoring systems. Understanding the importance of water quality and how to interpret data is essential for effective management.

5. Data Management: Develop a system for collecting, storing, and analyzing water quality data. Use software and applications that facilitate data visualization and trend analysis.

6. Continuous Improvement: Regularly review the performance of the monitoring system and make adjustments as needed. Stay updated with the latest advancements in technology and incorporate new tools that can enhance water quality management.

Case Studies and Examples

1. Case Study 1: Small-Scale Crawfish Farm

   - A small-scale crawfish farm implemented a combination of handheld digital meters and an automated water quality monitoring system. By regularly checking parameters such as DO and pH with handheld meters and using the automated system for continuous monitoring of temperature and ammonia levels, the farm achieved significant improvements in crawfish survival rates and growth performance.

2. Case Study 2: Commercial Crawfish Operation

   - A commercial crawfish operation integrated an IoT-based monitoring system that included sensors for DO, pH, temperature, and ammonia. The system provided real-time data and alerts, allowing the farm to quickly respond to changes in water quality. The use of machine learning algorithms helped optimize feeding schedules and aeration, resulting in increased productivity and reduced operational costs.

Future Trends and Innovations

The field of water quality monitoring in aquaculture is continually evolving. Future trends and innovations are likely to include:

1. Advanced Sensor Technologies: Development of more sensitive and accurate sensors that can detect a wider range of parameters at lower concentrations.

2. Integration with Mobile Technology: Enhanced mobile applications that allow farmers to monitor water quality and manage farm operations from their smartphones or tablets.

3. Blockchain for Data Security: Using blockchain technology to ensure the security and transparency of water quality data, providing traceability and trust in the aquaculture supply chain.

4. Sustainable Practices: Increasing focus on sustainable water management practices, including the use of biofilters and natural water treatment systems to maintain water quality.

In conclusion, advanced water quality monitoring technologies are indispensable for modern crawfish farming. They provide the tools and insights necessary to maintain optimal water conditions, ensuring the health and productivity of the crawfish. By embracing these technologies, farmers can achieve greater efficiency, sustainability, and profitability in their operations.

PART IX
Common Challenges and Solutions

9.1 Dealing with Pests and Predators

9.1.1 Common Pests

In the realm of crawfish farming, maintaining a healthy environment is paramount to ensuring the success and sustainability of your operation. Pests can pose significant challenges, affecting not only the crawfish population but also the overall health of the aquatic ecosystem. This section delves into the common pests encountered in crawfish farming and offers insights into their identification and impact.

1. Insects:

One of the most prevalent pests in crawfish ponds is insects. Various types of insects can become problematic, particularly those that thrive in aquatic environments.

- Mosquitoes: These are perhaps the most common insect pests in crawfish ponds. They breed in stagnant water, and their larvae can compete with crawfish for food. Moreover, adult mosquitoes can be a nuisance to both the crawfish and the farm workers.

- Water Boatmen and Backswimmers: These insects are often found in crawfish ponds. Water boatmen feed on algae and detritus, which can reduce food availability for crawfish. Backswimmers, on the other hand, are predatory and may feed on small crawfish, affecting the population dynamics.

2. Aquatic Plants:

Certain aquatic plants can also be considered pests due to their invasive nature and competition for resources.

- Duckweed: This small, free-floating plant can quickly cover the surface of a pond, blocking sunlight and reducing oxygen levels. This can create a stressful environment for crawfish, inhibiting their growth and reproduction.

- Water Hyacinth: Known for its rapid growth, water hyacinth can choke waterways and ponds, leading to decreased water quality and hindered crawfish movement.

3. Predatory Fish:

Predatory fish can pose a significant threat to crawfish populations. These fish can be either introduced accidentally or may enter the ponds from nearby water bodies.

- Bass and Catfish: Both of these species are known to feed on crawfish. Their presence in a crawfish pond can lead to substantial losses, particularly if they breed and establish a population within the pond.

4. Birds:

Birds can be major predators of crawfish, particularly those species that are adept at hunting in aquatic environments.

- Herons and Egrets: These wading birds are efficient hunters and can consume large numbers of crawfish, especially during the early morning and late afternoon when they are most active.

- Kingfishers: These birds dive into the water to catch prey, including young crawfish. Their sharp beaks and keen eyesight make them formidable predators.

5. Mammals:

Several mammalian species are known to prey on crawfish or cause indirect harm to their habitat.

- Raccoons: These nocturnal animals are adept at foraging in shallow waters and can cause significant losses in crawfish populations. They are known to raid ponds, particularly during the breeding season.

- Otters: While less common, otters can be highly destructive due to their feeding habits. A single otter can consume a large number of crawfish in a short period.

6. Parasites:

Parasites can affect crawfish health and productivity, often leading to disease outbreaks.

- Parasitic Copepods: These tiny crustaceans can attach themselves to crawfish gills and body surfaces, causing irritation and hampering respiration.

- Nematodes: Some nematode species can infect crawfish, leading to decreased growth rates and increased mortality.

Impact of Common Pests:

The impact of pests on crawfish farming can be multifaceted, ranging from direct predation to competition for resources and spread of diseases. For instance, insects like mosquitoes can reduce the availability of food for crawfish larvae, leading to slower growth rates. Predatory fish, birds, and mammals directly reduce crawfish numbers through consumption. Aquatic plants like duckweed and water hyacinth can degrade water quality, causing stress and lowering the survival rates of crawfish.

Parasites such as copepods and nematodes can weaken crawfish, making them more susceptible to other diseases and environmental stresses. This, in turn, can lead to significant economic losses for farmers due to reduced yields and increased management costs.

Identification and Monitoring:

Effective management of pests begins with accurate identification and regular monitoring. Farmers should conduct routine inspections of their ponds, paying close attention to the presence of unusual plant growth, insect larvae, and signs of predation.

- Visual Inspection: Regular visual inspections during different times of the day can help in identifying the presence of birds and mammals. Look for signs like tracks, droppings, and damaged vegetation around the pond edges.

- Water Quality Testing: Monitoring water quality parameters such as oxygen levels, pH, and nutrient concentrations can help detect early signs of pest-related problems. For example, a sudden drop in oxygen levels might indicate excessive plant growth or decaying organic matter due to insect larvae.

- Trapping and Sampling: Using traps and nets can help in identifying and quantifying the presence of aquatic insects and predatory fish. Regular sampling of crawfish can also help detect parasitic infestations.

Integrated Pest Management (IPM):

An integrated pest management approach is essential for effectively dealing with common pests in crawfish farming. IPM involves a combination of cultural, biological, and chemical control methods tailored to specific pests and environmental conditions.

- Cultural Control: This involves practices that reduce the habitat suitability for pests. For example, maintaining proper water levels and flow can reduce mosquito breeding sites. Regularly clearing invasive aquatic plants can prevent them from dominating the pond environment.

- Biological Control: Introducing natural predators or beneficial organisms can help control pest populations. For instance, fish species that feed on mosquito larvae can be introduced to control mosquito populations. However, care must be taken to select species that do not pose a threat to crawfish.

- Chemical Control: While chemical control should be a last resort, it can be necessary in severe infestations. Pesticides and herbicides should be used judiciously, ensuring they are safe for crawfish and the broader aquatic ecosystem. It is important to follow all regulations and guidelines to minimize environmental impact.

Conclusion:

Dealing with common pests in crawfish farming requires a comprehensive understanding of the various pests and their impacts on the ecosystem. By employing regular monitoring and integrated pest management strategies, farmers can effectively mitigate the threats posed by insects, aquatic plants, predatory fish, birds, mammals, and parasites. Maintaining a balanced and healthy pond environment is key to ensuring the productivity and sustainability of crawfish farming operations.

9.1.2 Effective Control Measures

Crawfish farming presents numerous challenges, particularly from pests and predators. Implementing effective control measures is crucial to ensure a healthy and thriving crawfish population. This section outlines a comprehensive strategy for managing these threats, including biological controls, habitat management, physical barriers, and chemical interventions.

Biological Control Measures

Biological control involves using natural predators or pathogens to control pest populations. In the context of crawfish farming, this method can be both environmentally friendly and cost-effective.

1. Predatory Fish: Introducing predatory fish, such as bass or catfish, can help control populations of small, unwanted species that compete with crawfish for food or habitat. However, it is essential to balance the number of predatory fish to avoid them preying on the crawfish.

2. Birds of Prey: Encouraging the presence of birds of prey, such as hawks or owls, can reduce the number of small mammals and birds that prey on crawfish. Installing perches around the farming area can attract these birds.

3. Biological Pathogens: Certain bacteria and fungi can target and kill specific pests. For instance, Bacillus thuringiensis (Bt) is a bacterium that produces toxins harmful to insect larvae. Using Bt as a biological insecticide can effectively manage pest populations without harming the crawfish.

Habitat Management

Proper habitat management can create an environment that is less conducive to pests and predators while promoting a healthy crawfish population.

1. Vegetation Control: Managing aquatic vegetation is crucial as it can harbor pests and provide cover for predators. Regularly cutting back excess vegetation can reduce hiding spots for pests and make the habitat less attractive to predators.

2. Water Quality Management: Maintaining optimal water quality through regular monitoring and adjustments helps to reduce the risk of disease and pest infestations. This includes ensuring appropriate pH levels, oxygenation, and minimizing pollutants.

3. Pond Design and Maintenance: Designing ponds with steep banks can help deter predators such as raccoons and herons, which prefer shallow areas. Regular maintenance, such as removing debris and repairing any damage to pond structures, can also help keep pests and predators at bay.

Physical Barriers

Physical barriers can effectively protect crawfish from both terrestrial and aquatic predators.

1. Fencing: Installing sturdy, fine-mesh fencing around the perimeter of the crawfish ponds can prevent access by terrestrial predators like raccoons, otters, and other small mammals. The fence should be buried at least a foot underground to prevent animals from digging underneath.

2. Netting: Covering the ponds with netting can protect crawfish from avian predators. The netting should be strong enough to withstand weather conditions and sufficiently fine to prevent birds from reaching through.

3. Traps and Deterrents: Using traps can help capture and remove persistent predators. Additionally, deterrents such as motion-activated lights or sprinklers can scare away predators without harming them.

Chemical Interventions

Chemical control should be used as a last resort due to potential environmental impacts and the risk of harming non-target species, including the crawfish.

1. Pesticides: If pest infestations become unmanageable through other methods, specific pesticides designed for aquatic use can be employed. It is crucial to select products that are safe for use in water bodies and follow the manufacturer's guidelines to avoid contaminating the habitat.

2. Herbicides: For managing excessive aquatic vegetation, selective herbicides can be used to target unwanted plants. Care must be taken to use herbicides that do not harm crawfish and to apply them in a manner that minimizes environmental impact.

3. Integrated Pest Management (IPM): Combining chemical treatments with biological and physical control measures in an Integrated Pest Management approach can effectively reduce pest populations while minimizing the negative effects on the environment and non-target organisms.

Monitoring and Adaptation

Continuous monitoring and adaptation of control measures are essential to effectively manage pests and predators in crawfish farming.

1. Regular Inspections: Conducting regular inspections of the ponds and surrounding areas helps in early detection of pest and predator activity. This allows for timely intervention before the situation escalates.

2. Record Keeping: Keeping detailed records of pest and predator incidences, along with the control measures implemented and their outcomes, helps in understanding patterns and improving future strategies.

3. Adaptive Management: As new information and technologies become available, it is important to adapt and refine control measures. This can include trying new biological controls, improving habitat management practices, or using more effective physical barriers.

Community and Cooperative Approaches

Collaborating with neighboring farms and participating in community pest management programs can enhance the effectiveness of control measures.

1. Shared Resources: Pooling resources for purchasing traps, netting, or biological control agents can reduce costs and improve efficiency.

2. Information Exchange: Sharing information about successful strategies and current pest challenges with other farmers can help develop more effective control measures and prevent widespread infestations.

3. Coordinated Efforts: Coordinating pest control efforts across multiple farms can create a more significant impact, reducing the overall pest and predator population in the area.

Conclusion

Effective control measures for managing pests and predators in crawfish farming require a multi-faceted approach, combining biological controls, habitat management, physical barriers, and, when necessary, chemical interventions. Continuous monitoring, adaptive management, and community cooperation further enhance the success of these measures, ensuring a healthy and productive crawfish farming operation. By implementing these strategies, farmers can mitigate the impact of pests and predators, ultimately leading to higher yields and more sustainable farming practices.

9.2 Managing Diseases

Identifying diseases in crawfish is crucial for maintaining a healthy and productive aquaculture operation. Diseases can significantly impact crawfish populations, leading to reduced yields, increased mortality, and economic losses. Proper disease identification enables timely and effective intervention, minimizing the negative effects on the farm.

9.2.1 Disease Identification

Introduction to Crawfish Diseases

Crawfish are susceptible to a variety of diseases caused by bacteria, viruses, fungi, and parasites. These diseases can spread rapidly in aquaculture systems, especially under poor water quality conditions or high stocking densities. Understanding the common signs and symptoms of diseases is the first step in disease identification and management.

Common Signs of Disease in Crawfish

Diseased crawfish often exhibit a range of symptoms that can include:

1. Lethargy and Reduced Activity: Infected crawfish may become less active, spending more time hiding or resting at the bottom of the tank or pond.

2. Changes in Coloration: Diseased crawfish can show abnormal coloration, such as darkening or whitening of the exoskeleton.

3. Soft Shell Syndrome: A condition where the exoskeleton becomes soft and fragile, making the crawfish more susceptible to injury and secondary infections.

4. Visible Lesions or Spots: Physical signs like ulcers, black spots, or reddish lesions on the exoskeleton are indicators of disease.

5. Erratic Swimming: Diseased crawfish may swim erratically or exhibit unusual behavior such as flipping upside down.

6. Decreased Feeding: Infected crawfish often show a reduced appetite or complete loss of interest in food.

7. Increased Mortality: A sudden increase in deaths within the population can be a sign of an underlying disease outbreak.

Bacterial Diseases

1. Vibriosis: Caused by Vibrio bacteria, this disease leads to symptoms such as darkening of the exoskeleton, lethargy, and increased mortality. Vibrio infections are often associated with poor water quality and high organic load.

2. Aeromonas Infection: Aeromonas hydrophila and other species cause ulcers, hemorrhages, and septicemia. Infected crawfish may display red discoloration on the body and limbs.

3. Shell Disease: Also known as Black Spot Disease, this condition is caused by bacteria such as Vibrio and Aeromonas. It results in dark, melanized spots on the exoskeleton and can lead to secondary fungal infections.

Viral Diseases

1. White Spot Syndrome Virus (WSSV): A highly contagious virus that affects many crustaceans, including crawfish. Infected crawfish exhibit white spots on the exoskeleton, lethargy, and high mortality rates.

2. Taura Syndrome Virus (TSV): Though more common in shrimp, TSV can infect crawfish, causing reddish discoloration, soft shell syndrome, and increased mortality.

Fungal Diseases

1. Fusarium spp.: Fungal infections can cause shell lesions, soft shell syndrome, and discoloration. Fusarium fungi thrive in environments with high organic matter and poor water quality.

2. Aphanomyces astaci (Crayfish Plague): This devastating fungal pathogen causes massive mortality in crawfish populations. Symptoms include lethargy, lack of coordination, and widespread death within a short period.

Parasitic Diseases

1. Microsporidia: These intracellular parasites can cause slow growth, poor feed conversion, and increased mortality. Infected crawfish may show muscle wasting and lethargy.

2. Branchiobdellid Worms: These ectoparasitic worms attach to the gills and other body parts, causing irritation, reduced growth, and secondary infections.

Diagnostic Techniques

Accurate diagnosis of diseases in crawfish involves a combination of visual inspection, microscopic examination, and laboratory testing.

1. Visual Inspection: Regular monitoring of the crawfish for signs of disease is essential. Look for changes in behavior, coloration, and physical condition.

2. Microscopic Examination: Collect samples from affected crawfish for microscopic analysis. This can help identify parasites, fungal spores, and bacterial colonies.

3. Laboratory Testing: Advanced diagnostic methods such as PCR (Polymerase Chain Reaction), histopathology, and bacteriological culture can confirm the presence of specific pathogens. These tests are often conducted by specialized laboratories.

Preventive Measures

Preventing disease outbreaks is more effective and economical than treating them after they occur. Key preventive measures include:

1. Water Quality Management: Maintain optimal water quality parameters, including temperature, pH, dissolved oxygen, and ammonia levels. Regular water changes and aeration can help achieve this.

2. Biosecurity Practices: Implement strict biosecurity measures to prevent the introduction and spread of pathogens. This includes quarantining new stock, disinfecting equipment, and controlling access to the farm.

3. Stocking Density Management: Avoid overcrowding to reduce stress and the risk of disease transmission. Follow recommended stocking densities for your specific crawfish species and farming system.

4. Healthy Nutrition: Provide a balanced diet with appropriate vitamins and minerals to boost the immune system of the crawfish. Avoid overfeeding, which can lead to water quality issues and increased disease risk.

5. Regular Health Monitoring: Conduct regular health checks and monitor for early signs of disease. Early detection allows for prompt intervention and reduces the impact of disease outbreaks.

Treatment Options

When a disease outbreak occurs, effective treatment is essential to control its spread and minimize losses. Treatment options vary depending on the type of pathogen involved.

1. Antibiotics: Bacterial infections can be treated with antibiotics such as oxytetracycline or sulfamerazine. However, the use of antibiotics should be carefully managed to avoid resistance and comply with regulations.

2. Antifungal Agents: Fungal infections can be treated with antifungal agents like malachite green or formalin. Proper dosage and application are crucial to avoid harming the crawfish or the environment.

3. Anti-Parasitic Treatments: Parasitic infections can be controlled with treatments such as formalin baths or potassium permanganate. Ensure proper handling and dosage to avoid toxicity.

4. Supportive Care: Provide supportive care to affected crawfish by improving water quality, reducing stress, and enhancing nutrition. This can help the crawfish recover more quickly and reduce mortality rates.

5. Vaccination: While still in the experimental stage for many aquaculture species, vaccines are being developed for some viral and bacterial diseases. Vaccination could offer a long-term solution to disease prevention.

Conclusion

Effective disease management in crawfish farming involves a combination of prevention, early detection, and timely intervention. By understanding the common diseases that affect crawfish, recognizing their symptoms, and implementing appropriate preventive and treatment measures, farmers can maintain healthy and productive crawfish populations. Continuous monitoring and adherence to best management practices are essential to minimize the impact of diseases and ensure the sustainability of the farming operation.

9.2.2 Treatment Options

Managing diseases in crawfish farming is a critical aspect of ensuring a healthy and productive crop. Disease outbreaks can lead to significant losses, making it essential to have effective treatment options. This section will explore various treatment strategies, ranging from chemical treatments to natural and preventive measures. By understanding these options, farmers can better protect their crawfish populations and maintain a successful farming operation.

Chemical Treatments

Chemical treatments are often the first line of defense against diseases in crawfish farming. These treatments involve the use of antibiotics, disinfectants, and other chemical agents to control and eradicate disease-causing pathogens.

1. Antibiotics: Antibiotics are commonly used to treat bacterial infections in crawfish. Tetracycline and oxytetracycline are examples of antibiotics that are effective against a range of bacterial diseases. These antibiotics can be administered through medicated feed or directly into the water. However, the use of antibiotics should be carefully managed to avoid the development of antibiotic-resistant bacteria.

2. Disinfectants: Disinfectants such as formalin, hydrogen peroxide, and iodine compounds are used to sanitize the farming environment. These agents help to reduce the pathogen load in the water and on the surfaces of equipment and structures. Regular disinfection protocols are essential to prevent the spread of diseases and maintain a healthy farming environment.

3. Chemical Bath Treatments: In some cases, individual crawfish or batches of crawfish can be treated with chemical baths. This method involves immersing the crawfish in a solution of a specific chemical agent for a defined period. Potassium permanganate and copper sulfate are examples of chemicals used in bath treatments. These treatments can be effective in controlling external parasites and infections.

Natural and Biological Treatments

Natural and biological treatments offer an alternative to chemical treatments and can be less harmful to the environment and the crawfish themselves. These methods focus on enhancing the natural defenses of the crawfish and utilizing beneficial organisms to control disease.

1. Probiotics: Probiotics are beneficial bacteria that can be added to the crawfish farming environment to improve the health and immunity of the crawfish. These bacteria compete with harmful pathogens for resources and space, thereby reducing the likelihood of disease outbreaks. Probiotics can be administered through feed or directly into the water.

2. Herbal Treatments: Various herbs and plant extracts have been shown to possess antimicrobial properties and can be used to treat diseases in crawfish. Garlic, neem, and turmeric are examples of plants with natural antibacterial and antifungal properties. These herbal treatments can be added to the crawfish feed or used to prepare bath treatments.

3. Biocontrol Agents: Biocontrol agents such as certain species of bacteria, fungi, and algae can be introduced into the farming environment to control harmful pathogens. These agents work by producing substances that inhibit the growth of pathogens or by directly attacking and consuming the pathogens. For example, the bacterium Bacillus subtilis is known for its ability to suppress the growth of harmful bacteria and fungi.

Environmental Management

Maintaining optimal environmental conditions is crucial for preventing and controlling diseases in crawfish farming. Proper management of water quality, temperature, and other environmental factors can significantly reduce the risk of disease outbreaks.

1. Water Quality Management: Regular monitoring and management of water quality parameters such as pH, dissolved oxygen, ammonia, and nitrite levels are essential. Poor

water quality can stress crawfish and make them more susceptible to diseases. Water quality can be improved through regular water changes, the use of aeration systems, and the addition of water conditioners.

2. Temperature Control: Crawfish are sensitive to temperature fluctuations, and extreme temperatures can lead to stress and increased susceptibility to diseases. Maintaining stable water temperatures within the optimal range for the specific species of crawfish being farmed is crucial. This can be achieved through the use of heaters, coolers, and shading structures.

3. Habitat Management: Providing a suitable habitat with adequate hiding places and proper substrate can help reduce stress and improve the overall health of crawfish. Dense vegetation, rocks, and artificial shelters can provide refuge and reduce aggressive interactions among crawfish.

Preventive Measures

Prevention is always better than cure. Implementing preventive measures can significantly reduce the risk of disease outbreaks and ensure the long-term health of the crawfish population.

1. Biosecurity Measures: Establishing strict biosecurity protocols is essential to prevent the introduction and spread of diseases. This includes controlling access to the farming site, disinfecting equipment and vehicles, and quarantining new stock before introducing them to the main population.

2. Health Monitoring: Regular health monitoring of the crawfish population can help detect early signs of disease and allow for prompt intervention. This involves visual inspections, sampling, and laboratory testing to identify pathogens and assess the overall health status of the crawfish.

3. Vaccination: Although vaccination is not yet widely used in crawfish farming, research is ongoing to develop vaccines against common diseases. Vaccination could provide a

proactive approach to disease management, enhancing the immune response of crawfish and reducing the need for chemical treatments.

Integrated Disease Management

An integrated approach to disease management combines multiple strategies to achieve the best results. This involves the use of chemical treatments, natural and biological methods, environmental management, and preventive measures in a coordinated manner.

1. Combining Treatments: In some cases, a combination of chemical and natural treatments may be more effective than using a single method. For example, antibiotics can be used to quickly reduce bacterial infections, while probiotics and herbal treatments can help restore and maintain a healthy microbial balance.

2. Rotational Use of Treatments: Rotating the use of different treatments can help prevent the development of resistant pathogens. By varying the treatments used, farmers can reduce the selective pressure on pathogens and maintain the effectiveness of available treatment options.

3. Continuous Improvement: Disease management strategies should be regularly reviewed and updated based on the latest research and field experiences. Farmers should stay informed about new treatment options, emerging diseases, and best practices in crawfish farming.

In conclusion, effective disease management in crawfish farming requires a multifaceted approach that includes chemical treatments, natural and biological methods, environmental management, and preventive measures. By implementing a comprehensive disease management plan, farmers can protect their crawfish populations, improve productivity, and ensure the sustainability of their farming operations.

PART IX: COMMON CHALLENGES AND SOLUTIONS

9.3 Environmental Challenges

The cultivation of crawfish, like any other form of aquaculture, is subject to various environmental challenges. These challenges not only impact the health and yield of the crawfish but also pose significant risks to the sustainability and profitability of farming operations. One of the most pressing environmental challenges is the impact of climate change. This section will delve into the various aspects of how climate change affects crawfish farming and the broader ecosystem.

9.3.1 Climate Change Impacts

Climate change is an overarching environmental issue that affects ecosystems worldwide, and crawfish farming is no exception. The impacts of climate change on crawfish farming can be multifaceted and complex, involving changes in temperature, precipitation patterns, water quality, and the frequency of extreme weather events. Understanding these impacts is crucial for developing adaptive strategies to ensure the sustainability of crawfish farming operations.

1. Temperature Fluctuations

One of the most direct effects of climate change is the alteration of temperature regimes. Crawfish are ectothermic animals, meaning their body temperature and metabolic processes are influenced by the ambient temperature. Optimal temperature ranges are essential for the growth, reproduction, and survival of crawfish.

- Growth and Metabolism: Increased temperatures can accelerate the metabolic rate of crawfish, leading to faster growth but also higher demand for oxygen and food. Conversely, temperatures that exceed the optimal range can cause stress, reduce growth rates, and increase mortality.

- Reproductive Cycles: Crawfish reproduction is closely tied to water temperature. Warmer temperatures can lead to earlier and more frequent breeding cycles. However, if the temperatures rise too high, it can disrupt reproductive processes, leading to reduced fertility and lower hatchling survival rates.

- Thermal Stress: Prolonged exposure to extreme temperatures can cause thermal stress, which weakens the immune system of crawfish, making them more susceptible to diseases and reducing their overall resilience.

2. Changes in Precipitation Patterns

Climate change can lead to significant changes in precipitation patterns, affecting both the quantity and distribution of rainfall. This has direct implications for water availability and quality in crawfish ponds.

- Water Levels: Increased rainfall can lead to higher water levels in ponds, potentially causing overflow and the loss of crawfish. On the other hand, prolonged droughts can reduce water levels, leading to overcrowding, increased competition for resources, and heightened stress among the crawfish population.

- Water Quality: Variability in rainfall can affect water quality through changes in salinity, pH, and the introduction of pollutants. Heavy rainfall can cause runoff that introduces pesticides, herbicides, and other contaminants into crawfish ponds, negatively impacting the health of the crawfish.

- Ecosystem Balance: Altered precipitation patterns can disrupt the balance of the pond ecosystem, affecting the availability of natural food sources for crawfish and altering the habitat conditions necessary for their growth and reproduction.

3. Frequency of Extreme Weather Events

Climate change is associated with an increase in the frequency and intensity of extreme weather events, such as hurricanes, floods, and heatwaves. These events can have devastating effects on crawfish farming.

- Storm Damage: Hurricanes and floods can cause significant physical damage to crawfish ponds, infrastructure, and equipment. They can also lead to the loss of crawfish through overflow and washout.

- Temperature Extremes: Heatwaves can lead to rapid increases in water temperature, causing thermal stress and potentially leading to mass die-offs. Conversely, unexpected cold snaps can also be detrimental, particularly if they occur during the breeding season or early development stages.

- Recovery and Resilience: The increased frequency of extreme weather events necessitates the development of robust recovery and resilience strategies. This includes reinforcing pond infrastructure, implementing early warning systems, and developing contingency plans to mitigate the impact of such events.

4. Ocean Acidification

Although primarily a concern for marine environments, ocean acidification can indirectly affect freshwater systems through changes in atmospheric CO_2 levels and the deposition of acidic compounds.

- Water Chemistry: Increased CO_2 levels can lead to the acidification of freshwater bodies, altering the water chemistry and impacting the availability of essential minerals and nutrients required for crawfish health.

- Physiological Stress: Acidic conditions can cause physiological stress in crawfish, affecting their growth, reproduction, and immune function. This stress can make crawfish more susceptible to diseases and reduce their overall survival rates.

5. Ecosystem Interactions

Climate change can alter the interactions between crawfish and other organisms in their ecosystem. This includes changes in the abundance and behavior of predators, prey, and competitors.

- Predator-Prey Dynamics: Changes in temperature and habitat conditions can affect the population dynamics and behavior of predators and prey. For example, warmer temperatures may lead to an increase in predator populations, putting additional pressure on crawfish.

- Invasive Species: Climate change can facilitate the spread of invasive species that compete with crawfish for resources or prey upon them. These invasive species can disrupt the balance of the pond ecosystem and pose significant challenges to crawfish farmers.

6. Impacts on Human Management Practices

Climate change can also affect the human dimension of crawfish farming, including changes in management practices, economic viability, and regulatory frameworks.

- Adaptation Strategies: Farmers may need to adopt new management practices to cope with the impacts of climate change. This includes changes in feeding regimes, pond management, and the use of climate-resilient crawfish strains.

- Economic Considerations: The increased costs associated with adapting to climate change, such as upgrading infrastructure and implementing mitigation measures, can affect the economic viability of crawfish farming. Farmers may need to explore new markets or diversify their income sources to remain profitable.

- Regulatory Changes: Governments and regulatory bodies may introduce new policies and regulations in response to climate change. These regulations could impact water use, environmental protection, and sustainable farming practices, requiring farmers to stay informed and comply with new standards.

Conclusion

Climate change presents a significant and multifaceted challenge to crawfish farming. The impacts of temperature fluctuations, changes in precipitation patterns, extreme weather events, ocean acidification, ecosystem interactions, and human management practices all contribute to the complexity of this issue. To ensure the sustainability and profitability of crawfish farming in the face of climate change, farmers must adopt adaptive strategies, invest in resilient infrastructure, and stay informed about emerging trends and regulatory changes. By proactively addressing these challenges, the crawfish farming industry can continue to thrive and contribute to food security and economic development in the regions where it is practiced.

9.3.2 Mitigation Strategies

Mitigating environmental challenges in crawfish farming involves implementing a combination of proactive and reactive measures to ensure the sustainability and productivity of crawfish ponds. The strategies encompass managing water quality, adapting to climate change, ensuring biodiversity, and implementing effective farming practices. This section provides an in-depth guide to various mitigation strategies that can help farmers address environmental challenges.

1. Water Quality Management

1.1 Regular Monitoring:

Maintaining optimal water quality is crucial for crawfish health and growth. Regularly test water parameters such as pH, dissolved oxygen, temperature, ammonia, nitrites, and nitrates. Utilize portable water testing kits or invest in automated monitoring systems to keep track of these parameters consistently.

1.2 Aeration:

Proper aeration ensures sufficient oxygen levels in the water. Install aerators or paddle wheels to circulate water and enhance oxygen distribution. This helps prevent hypoxic conditions, which can stress or kill crawfish.

1.3 Filtration Systems:

Implementing filtration systems can help remove excess nutrients, waste products, and harmful chemicals from the water. Biological filters, which use beneficial bacteria to break down waste, can be particularly effective in maintaining water quality.

1.4 Buffering Agents:

Use buffering agents to stabilize pH levels in the pond. Crushed limestone or agricultural lime can help maintain a pH range of 7.0 to 8.5, which is ideal for crawfish.

2. Climate Change Adaptation

2.1 Pond Design and Location:

Design ponds with varying depths to provide refuges for crawfish during extreme weather conditions. Consider constructing ponds in areas with natural shade or planting vegetation around ponds to reduce temperature fluctuations.

2.2 Seasonal Adjustments:

Adapt farming practices to seasonal changes. For example, during hotter months, increase water depth to mitigate high temperatures and evaporation. In cooler months, ensure that ponds do not freeze by maintaining adequate water flow.

2.3 Shade Structures:

Install shade structures or floating covers to reduce direct sunlight exposure, which can increase water temperature. Shade structures can also reduce algae blooms by limiting the amount of sunlight penetrating the water.

2.4 Water Storage:

Develop water storage systems to collect and store rainwater. This stored water can be used during dry spells to maintain pond levels and reduce reliance on external water sources.

3. Biodiversity and Habitat Management

3.1 Vegetation Management:

Promote the growth of beneficial aquatic plants that provide habitat and food for crawfish. Plants like water lilies and cattails can offer shelter and help maintain water quality by absorbing excess nutrients.

3.2 Predator Control:

Implement measures to control predators such as birds, raccoons, and fish that prey on crawfish. Use nets, fences, or scare devices to protect ponds. Additionally, maintain a balanced ecosystem by controlling invasive species that can disrupt the habitat.

3.3 Integrated Pest Management (IPM):

Develop an IPM plan to manage pests in a sustainable way. This can include biological controls, such as introducing natural predators or competitors, and mechanical controls, like traps and barriers.

4. Sustainable Farming Practices

4.1 Rotational Harvesting:

Practice rotational harvesting to avoid overexploitation of the pond. This ensures that there is always a breeding population of crawfish in the pond, promoting sustainability.

4.2 Polyculture:

Consider integrating polyculture systems, where crawfish are farmed alongside other compatible species such as fish or certain plants. This can enhance productivity and resource use efficiency.

4.3 Organic Farming:

Adopt organic farming practices by avoiding the use of synthetic chemicals and fertilizers. Instead, use organic alternatives such as compost or organic pellets that are environmentally friendly and safe for crawfish.

4.4 Waste Management:

Implement effective waste management practices to prevent the accumulation of organic waste, which can deteriorate water quality. Regularly remove uneaten feed and debris from the pond. Composting organic waste can also be a sustainable practice.

5. Technological Innovations

5.1 Automated Feeding Systems:

Use automated feeding systems to ensure precise and consistent feeding schedules. This reduces feed waste and ensures that crawfish receive adequate nutrition.

5.2 Remote Sensing and Monitoring:

Invest in remote sensing and monitoring technologies to keep track of environmental parameters in real-time. These technologies can provide early warnings of adverse conditions, allowing for timely interventions.

5.3 Data Analytics:

Utilize data analytics to analyze trends and patterns in environmental conditions and crawfish health. This can help in making informed decisions and improving farm management practices.

6. Community and Policy Engagement

6.1 Collaborative Efforts:

Engage with local farming communities to share knowledge and resources. Collaborative efforts can lead to better management practices and collective problem-solving.

6.2 Policy Advocacy:

Work with local and national authorities to advocate for policies that support sustainable crawfish farming. This can include regulations on water usage, environmental protection measures, and support for research and development.

6.3 Education and Training:

Invest in education and training programs for farmers and farm workers. Providing knowledge on best practices and new technologies can enhance the overall productivity and sustainability of the farm.

Conclusion

Mitigating environmental challenges in crawfish farming requires a comprehensive and adaptive approach. By implementing strategies such as water quality management, climate change adaptation, habitat management, sustainable farming practices, and technological innovations, farmers can ensure the long-term viability and productivity of their operations. Collaborative efforts and policy engagement further support these initiatives, fostering a sustainable future for crawfish farming.

CHAPTER X
Appendices

10.1 Glossary of Terms

Aeration

The process of adding oxygen to water, which is crucial for the survival and growth of crawfish. Aeration can be achieved through mechanical devices such as aerators, fountains, or diffusers.

Aquaculture

The farming of aquatic organisms such as fish, crustaceans, mollusks, and aquatic plants. In this context, it specifically refers to the farming of crawfish.

Bait

Food or attractants used to lure crawfish into traps. Common baits include fish parts, chicken, or commercial crawfish bait.

Biosecurity

Measures and protocols implemented to protect the crawfish population from diseases, pests, and invasive species. This includes cleaning equipment, controlling access to ponds, and monitoring water quality.

Broodstock

Mature crawfish used for breeding purposes. The selection of healthy broodstock is essential for producing a robust and resilient crawfish population.

Buffer Zone

A protective area surrounding a crawfish pond designed to prevent contamination from external sources such as pesticides, herbicides, and runoff.

Catch Per Unit Effort (CPUE)

A measure used to assess the efficiency of crawfish trapping. It is calculated by dividing the number of crawfish caught by the number of traps used or the amount of time spent trapping.

Chemical Treatment

The use of chemicals to control unwanted pests, diseases, or vegetation in crawfish ponds. Proper usage and regulations must be followed to avoid harming the crawfish and the environment.

Chitin

A complex carbohydrate that forms the exoskeleton of crawfish. Chitin is important for the structural integrity and protection of crawfish.

Crawfish Trap

A device used to catch crawfish, typically made of wire mesh with one or more funnel-shaped entrances. Traps are baited to attract crawfish and placed in ponds or streams.

Decapod

A classification of crustaceans that includes crawfish, characterized by having ten legs. Crawfish are decapods and share common features with shrimp, crabs, and lobsters.

Detritus

Organic matter produced by the decomposition of plants and animals. Detritus serves as an important food source for crawfish, providing nutrients necessary for their growth.

Dissolved Oxygen (DO)

The amount of oxygen present in water, which is essential for the respiration of aquatic organisms. Maintaining adequate DO levels is crucial for the health of crawfish.

Ecdysis

The process of molting in crawfish, where they shed their old exoskeleton to grow a new, larger one. This process is vital for growth and development but also makes them vulnerable to predation.

Ecosystem

A biological community of interacting organisms and their physical environment. Crawfish farming aims to create a balanced ecosystem within the pond to support healthy crawfish populations.

Feed Conversion Ratio (FCR)

A measure of the efficiency with which crawfish convert feed into body mass. A lower FCR indicates more efficient feed usage, which is desirable for cost-effective farming.

Fingerlings

Young crawfish that have recently hatched and are in the early stages of development. Proper care of fingerlings is crucial for ensuring a healthy and productive adult population.

Habitat

The natural environment in which crawfish live, including the water quality, vegetation, and substrate. Creating and maintaining suitable habitat conditions is essential for successful crawfish farming.

Harvest

The process of collecting mature crawfish from ponds or traps for sale or consumption. Timing and techniques of harvest can impact the quality and quantity of the crawfish yield.

Hatchery

A facility where crawfish eggs are incubated and hatched under controlled conditions. Hatcheries play a critical role in supplying young crawfish (fingerlings) for stocking ponds.

Hydrology

The study of the properties and distribution of water on Earth. Understanding hydrology is important for managing water resources in crawfish farming, including pond construction and water management.

Invasive Species

Non-native species that can cause harm to the environment, economy, or human health. Invasive species can compete with crawfish for resources and introduce diseases.

Larvae

The early developmental stage of crawfish following hatching. Larvae undergo several molts before becoming juvenile crawfish.

Liming

The application of lime to ponds to adjust pH levels and improve water quality. Proper liming practices help create a suitable environment for crawfish growth and survival.

Molting

The shedding of the exoskeleton in crawfish, allowing for growth. Molting is a critical phase in the life cycle of crawfish, requiring specific environmental conditions.

Nursery Pond

A pond used to rear young crawfish until they reach a size suitable for transfer to grow-out ponds or for sale. Nursery ponds provide a controlled environment for the early stages of crawfish development.

Omnivorous

An organism that eats both plant and animal matter. Crawfish are omnivorous, feeding on a variety of food sources including detritus, algae, plants, and small aquatic animals.

Organic Matter

Material derived from living organisms. Organic matter in ponds, such as decomposing plant material, provides nutrients and habitat for crawfish.

pH

A measure of the acidity or alkalinity of water. Maintaining the proper pH balance is important for crawfish health and can affect their growth and reproduction.

Plankton

Small and microscopic organisms that float or drift in water. Plankton serves as a food source for young crawfish and contributes to the overall productivity of the pond ecosystem.

Polyculture

The practice of raising more than one species in the same pond. In crawfish farming, polyculture can involve raising crawfish alongside fish or other aquatic organisms to enhance productivity and resource use.

Population Density

The number of crawfish per unit area in a pond. Managing population density is crucial to avoid overcrowding, which can lead to poor growth and increased susceptibility to disease.

Predation

The act of one organism feeding on another. In crawfish farming, predation can be a significant challenge, requiring measures to protect crawfish from predators such as birds, fish, and other animals.

Reproduction Cycle

The sequence of events from mating to the release of eggs and the hatching of larvae. Understanding the reproduction cycle is essential for timing breeding and ensuring a steady supply of young crawfish.

Salinity

The concentration of salts in water. Crawfish generally prefer freshwater, so monitoring and managing salinity levels is important in crawfish farming.

Sedimentation

The process by which particles settle out of water. Sedimentation can affect water quality and the health of crawfish, requiring regular pond maintenance to manage sediment levels.

Stocking

The introduction of young crawfish into ponds to establish or replenish populations. Stocking rates and methods can impact the success of a crawfish farming operation.

Substrate

The bottom surface of a pond, which can include soil, sand, gravel, or organic material. The type and condition of the substrate influence the habitat suitability for crawfish.

Sustainable Farming

Practices that maintain or enhance the quality of the environment while producing food. In crawfish farming, sustainability involves managing resources to ensure long-term viability and minimal environmental impact.

Temperature

A key environmental factor affecting the growth, reproduction, and survival of crawfish. Monitoring and managing water temperature is essential for successful crawfish farming.

Trap Effort

The amount of effort expended in setting and retrieving traps. Efficient trap effort maximizes the catch and reduces labor costs in crawfish farming.

Water Quality

The condition of water, including factors such as temperature, pH, dissolved oxygen, and clarity. Good water quality is critical for the health and productivity of crawfish.

Yield

The total quantity of crawfish produced or harvested from a pond. Yield is a primary measure of success in crawfish farming, influenced by various factors including management practices, water quality, and stocking rates.

PART X: APPENDICES

10.2 Sample Budget Templates

Creating an effective budget is crucial for the success of any crawfish farming venture. A well-structured budget helps in planning, resource allocation, cost management, and financial forecasting. In this section, we provide comprehensive sample budget templates to guide you through the process. These templates cover various aspects of crawfish farming, from initial setup costs to ongoing operational expenses.

Template 1: Initial Setup Budget

Purpose: This template is designed to help you estimate the costs involved in setting up a crawfish farm. It includes expenses related to land acquisition, construction of ponds, equipment purchase, and initial stocking of crawfish.

Item	Description	Estimated Cost (USD)
Land Acquisition	Purchase or lease of land suitable for ponds	$5,000 - $20,000
Pond Construction	Excavation, lining, and initial water setup	$10,000 - $50,000
Water Supply System	Pumps, pipes, and irrigation setup	$2,000 - $10,000
Fencing	Installation of fencing around the property	$1,000 - $3,000
Hatchery Setup	Tanks, filtration, and breeding equipment	$3,000 - $10,000
Initial Stock	Purchase of juvenile crawfish	$1,000 - $5,000
Equipment	Aerators, feed dispensers, and maintenance tools	$2,000 - $8,000
Licensing and Permits	Legal requirements and regulatory fees	$500 - $1,500
Miscellaneous	Unexpected expenses	$1,000 - $3,000
Total Estimated Cost		**$25,500 - $110,500**

Template 2: Annual Operational Budget

Purpose: This template helps you plan for the annual operating costs of running a crawfish farm. It includes regular expenses such as feed, labor, maintenance, utilities, and marketing.

Item	Description	Estimated Cost (USD)
Feed	Crawfish feed and supplements	$3,000 - $10,000
Labor	Salaries for farm workers and management	$20,000 - $50,000
Utilities	Electricity, water, and other utilities	$2,000 - $5,000
Maintenance	Pond repairs, equipment maintenance	$1,500 - $5,000
Health Management	Disease prevention, veterinary services	$1,000 - $3,000
Transportation	Fuel, vehicle maintenance, and transportation costs	$2,000 - $5,000
Marketing	Advertising, promotions, and sales activities	$1,000 - $3,000
Insurance	Property and liability insurance	$1,500 - $3,000
Miscellaneous	Unforeseen operational expenses	$1,000 - $3,000
Total Estimated Cost		**$33,000 - $87,000**

Template 3: Monthly Cash Flow Budget

Purpose: To monitor the monthly inflow and outflow of cash. This template helps in maintaining liquidity and ensuring that the farm can meet its financial obligations each month.

PART X: APPENDICES

Month	Estimated Income (USD)	Estimated Expenses (USD)	Net Cash Flow (USD)
January	$4,000	$5,500	-$1,500
February	$4,500	$5,000	-$500
March	$5,000	$5,500	-$500
April	$6,000	$6,000	$0
May	$7,000	$5,500	$1,500
June	$8,000	$6,000	$2,000
July	$9,000	$5,500	$3,500
August	$10,000	$5,000	$5,000
September	$8,000	$5,500	$2,500
October	$7,000	$6,000	$1,000
November	$6,000	$5,500	$500
December	$5,000	$5,000	$0
Total	$79,500	$66,000	$13,500

Template 4: Profit and Loss Statement

Purpose: To summarize the revenues, costs, and expenses incurred during a specific period, providing a snapshot of the farm's financial performance.

PART X: APPENDICES

Item	Description	Amount (USD)
Revenue		
Crawfish Sales	Income from selling crawfish	$90,000
Other Income	Income from by-products or services	$5,000
Total Revenue		$95,000
Cost of Goods Sold		
Feed	Costs associated with crawfish feed	$10,000
Labor	Wages and salaries for workers	$40,000
Maintenance	Repairs and maintenance of equipment and ponds	$5,000
Health Management	Disease prevention and veterinary costs	$3,000
Utilities	Water, electricity, and other utilities	$5,000
Total COGS		$63,000
Gross Profit		$32,000
Operating Expenses		
Marketing	Advertising and promotional expenses	$3,000
Transportation	Fuel and vehicle maintenance costs	$4,000
Insurance	Property and liability insurance	$3,000
Miscellaneous	Other operating expenses	$2,000
Total Operating Expenses		$12,000
Net Profit		$20,000

Template 5: Break-Even Analysis

Purpose: To determine the level of sales necessary to cover all costs, providing insights into the financial viability and the minimum performance required to avoid losses.

Item	Description	Amount (USD)
Fixed Costs		
Land Lease or Mortgage	Cost of land lease or mortgage payments	$5,000
Salaries	Fixed salaries for permanent staff	$20,000
Insurance	Annual insurance costs	$3,000
Utilities	Fixed utility costs	$2,000
Maintenance	Fixed maintenance costs	$2,000
Total Fixed Costs		**$32,000**
Variable Costs per Unit		
Feed	Cost of feed per unit of crawfish produced	$1
Labor	Cost of labor per unit of crawfish produced	$2
Utilities	Variable utility cost per unit	$0.50
Health Management	Health and veterinary costs per unit	$0.50
Total Variable Costs/Unit		**$4**
Sales Price per Unit	Average selling price per unit of crawfish	$10
Break-Even Point (Units)	Fixed Costs / (Sales Price - Variable Costs)	32,000 / (10 - 4) = 5,334 units

Template 6: Capital Expenditure Plan

Purpose: To outline the planned investments in long-term assets such as land, buildings, and equipment, providing a roadmap for capital allocation over a specified period.

Item	Description	Year 1 (USD)	Year 2 (USD)	Year 3 (USD)
Land Acquisition	Purchase of additional land for expansion	$10,000	$0	$0
Pond Construction	Building new ponds	$20,000	$15,000	$10,000
Equipment Purchase	New aerators, pumps, and maintenance tools	$8,000	$5,000	$5,000
Hatchery Expansion	Adding more tanks and filtration systems	$5,000	$10,000	$5,000
Technology Upgrades	Installing automated feeding and monitoring	$7,000	$7,000	$0
Vehicles	Purchase of transportation vehicles	$5,000	$0	$5,000
Total Capital Expenditure		$55,000	$37,000	$25,000

These sample budget templates serve as a foundational guide for managing the financial aspects of your crawfish farming business. It's essential to customize these templates based on your specific circumstances and continuously update them as your business evolves. Regular financial reviews and adjustments will ensure that you stay on track and achieve your business goals effectively.

CONCLUSION

Crawfish farming, a practice deeply rooted in tradition and continuously evolving with modern advancements, presents a unique and rewarding opportunity for both seasoned farmers and enthusiastic beginners. Throughout this guide, we have meticulously explored the fundamental aspects and nuanced details essential to successful crawfish farming. From understanding the biology and lifecycle of crawfish to setting up the perfect habitat, managing water quality, feeding, and finally, to harvesting, every chapter has been designed to equip you with the knowledge and tools necessary for a thriving crawfish farm.

Embarking on this journey requires dedication, patience, and a willingness to learn and adapt. The challenges you may face, such as fluctuating weather conditions, disease management, and maintaining optimal water quality, are all part of the process. However, the rewards—both in terms of the satisfaction of nurturing a thriving crawfish population and the potential financial gains—are well worth the effort.

By now, you should have a comprehensive understanding of the critical steps involved in crawfish farming. Whether you are planning to start small with a backyard pond or envisioning a larger commercial operation, the principles remain the same. Success lies in the details: regular monitoring, timely interventions, and a deep appreciation for the ecosystem you are cultivating.

In the ever-evolving landscape of aquaculture, staying informed and connected with the broader farming community is crucial. As you move forward, continue to seek out new information, stay updated on best practices, and network with fellow farmers. This guide is just the beginning; your real education will come from hands-on experience and continuous learning.

Acknowledgements

We are deeply grateful to our readers for choosing "Crawfish Farming for Beginners: Step-by-Step Guide." Your decision to invest in this book signifies a commitment to understanding and mastering the art of crawfish farming, and for that, we extend our heartfelt thanks.

Creating this guide has been a labor of love, driven by a passion for aquaculture and a desire to share valuable knowledge with others. We owe a debt of gratitude to the numerous experts and experienced farmers who generously shared their insights and expertise, making this book a comprehensive resource for crawfish enthusiasts.

To our families and friends, thank you for your unwavering support and encouragement throughout the writing process. Your belief in this project was a constant source of motivation.

Finally, to you, the reader: your interest in crawfish farming is the reason this book exists. We hope that the information and guidance provided within these pages will serve as a reliable foundation for your farming endeavors. As you embark on this exciting journey, remember that success in farming comes from patience, persistence, and a genuine love for the work you do. We wish you all the best in your crawfish farming adventure and look forward to hearing about your successes.

Thank you for trusting us to be part of your journey. Happy farming!

www.ingramcontent.com/pod-product-compliance
Lightning Source LLC
Chambersburg PA
CBHW062312220526
45479CB00004B/1140